AF388892

Advances in Object and Activity Detection in Remote Sensing Imagery

Advances in Object and Activity Detection in Remote Sensing Imagery

Editors

Anwaar Ulhaq
Douglas Pinto Sampaio Gomes

MDPI • Basel • Beijing • Wuhan • Barcelona • Belgrade • Manchester • Tokyo • Cluj • Tianjin

Editors

Anwaar Ulhaq
Charles Sturt University
Australia

Douglas Pinto Sampaio Gomes
Charles Sturt University
Australia

Editorial Office
MDPI
St. Alban-Anlage 66
4052 Basel, Switzerland

This is a reprint of articles from the Special Issue published online in the open access journal *Remote Sensing* (ISSN 2072-4292) (available at: https://www.mdpi.com/journal/remotesensing/special_issues/Object_Activity_Detection).

For citation purposes, cite each article independently as indicated on the article page online and as indicated below:

LastName, A.A.; LastName, B.B.; LastName, C.C. Article Title. *Journal Name* **Year**, *Volume Number*, Page Range.

ISBN 978-3-0365-4229-4 (Hbk)
ISBN 978-3-0365-4230-0 (PDF)

Contents

About the Editors

Anwaar Ulhaq

Dr Anwaar Ulhaq is serving as a senior lecturer and deputy leader, Machine Vision and Digital Health Research in the School of Computing, Mathematics and Engineering, Charles Sturt University, Australia. Anwaar holds a PhD (Artificial Intelligence) from Monash University, Australia. He has completed professional education in machine learning and artificial intelligence from the Massachusetts Institute of Technology (MIT). He has extensive teaching and research experience from reputed Australian universities, including Victoria University, Swinburne University of Technology, and Central Queensland University. He has also worked as a research fellow at the Institute for Sustainable Industries and Liveable Cities, Victoria University, Australia. His research interests include artificial creativity, deep learning, data analytics, and computer vision. He has published more than 60 peer-reviewed papers in reputed journals and conferences.

Douglas Pinto Sampaio Gomes

Dr. Douglas has a Bachelor's, Master's (USP—Brazil), and Doctorate (Victoria University—Australia) in Electrical Engineering. Their past works involved applications in power distribution system protection such as optimization techniques for the deployment of monitoring devices and the application of machine learning for signal classification of faults. Dr. Douglas is published and has served as a reviewer in journals such as the *IEEE Transactions on Power Delivery* and *IEEE Transactions on Instrumentation and Measurement*. Their current works focus further on signal processing and machine learning with applications related to the fields of image classification, power system protection, and powerline communications. Recent papers contemplate the use of Deep Learning for tasks such as detecting and segmentation in medical and plant images, as well as the study of high-impedance faults in power distribution systems. At present, Dr. Douglas is employed as a Research Officer at Victoria University, Melbourne, working on building real-time devices for the detection of power system faults leading to fire ignition in vegetation. Building and deploying systems lead to experiences in developing hardware and software such as analog interfaces, signal processing in FPGAs, real-time services in digital signal processors (DSPs), and applications in microprocessors (ARM). Dr. Douglas's interests for collaboration remain in leveraging and deploying machine learning and signal processing solutions to devices addressing practical and impactful problems.

 *remote sensing*

Editorial

Editorial for the Special Issue "Advances in Object and Activity Detection in Remote Sensing Imagery"

Anwaar Ulhaq [1,2,*] **and Douglas Pinto Sampaio Gomes** [1]

1   School of Computing, Mathematics, and Engineering, Charles Sturt University,
    Port Macquarie, NSW 2444, Australia; dgomes@csu.edu.au
2   The Institute for Sustainable Industries and Liveable Cities (ISILC), College of Engineering and Science,
    Victoria University, Melbourne, VIC 8001, Australia
*   Correspondence: aulhaq@csu.edu.au

Citation: Ulhaq, A.; Gomes, D.P.S. Editorial for the Special Issue "Advances in Object and Activity Detection in Remote Sensing Imagery". *Remote Sens.* **2022**, *14*, 1844. https://doi.org/10.3390/rs14081844

Received: 7 April 2022
Accepted: 11 April 2022
Published: 12 April 2022

Publisher's Note: MDPI stays neutral with regard to jurisdictional claims in published maps and institutional affiliations.

Advances in data collection and accessibility, such as unmanned aerial vehicle (UAV) technology, the availability of satellite imagery, and the increasing performance of deep learning models, have had significant impacts on solving various remote sensing problems and proposing new applications ranging from vegetation and wildlife monitoring to crowd monitoring. This Special Issue contains seven high-quality papers [1–7] approaching problems relating to object detection, semantic segmentation, and multi-modal data alignment. In terms of the methods utilized, it is not surprising that six of the seven papers on this issue involve the application of deep learning. The papers also attest to the powerful aspect of the field where researchers can collaborate and validate their work on open-source models and datasets.

The first paper [1] addresses the problem of animal population estimation via thermal images, which often face the challenge of being low-resolution. The authors propose a modification to a popular object detection framework, naming it Distant-YOLO. The improved model, trained on a dataset containing low-resolution aerial images of rabbits, kangaroos, and pigs, was capable of detecting such animals, thus being potentially relevant for wildlife researchers and managers that previously relied on manual annotations.

The second paper [2] focuses on the detection of ships from synthetic-aperture radar images. The addressed problem relates to the fact that the accuracy of detection systems is often negatively affected by the complex background interference and the multi-scale features of ships. The authors propose the Quad-FPN architecture, a combination of four feature pyramid networks, which are all individually validated with extensive ablation studies. The work is potentially relevant for extremely important problems such as marine surveillance, traffic control, and fishery management. Likewise, the third paper [3] proposes an innovative architecture for the detection of objects through satellite optical imagery such as ships, but in a generalized manner. The authors address the problem that arbitrary objects in satellite imagery still pose a serious challenge for object detection models due to their diverse patterns in orientation, scale, and aspect ratio. The resulting model composed of such an active feature map realignment achieves higher performance, validated by the achievement of state-of-the-art results in two public datasets.

Semantic-segmentation-centered problems were also addressed with two papers presenting innovative enhancements. One semantic segmentation paper [4] aimed at urban vegetation cover estimation while addressing a problem often present in similar works given by an over-reliance on image color attributes. The improvement proposed is composed of a Multiview Semantic Vegetation Index (MSVI), which is implemented by a segmentation model (FCN and U-net) and with a proposed color mask adjustment. Given its multiview capability and ability to be applied to images such as those from Google panoramic cameras, the method has potential implications for real-time vegetation monitoring. In the second semantic segmentation paper [5], the subject approached was tidal

---

Remote Sens. **2022**, *14*, 1844

flat waterbody estimation. Such a problem suffers from particular challenges where waterbodies differ little between their background while also contemplating blurry boundaries, which are difficult to detect accurately. As such a task represents one of the main ways to estimate waterbodies, it is increasingly relevant to aspects such as ecosystem protection and restoration, pollution control, and infrastructure construction.

The other two papers proposing novel solutions addressed the problems of crowd estimation and multi-camera space alignment. The innovative solution, presented by the paper [6] on crowd estimation by UAVs, aims at improving challenges in the form of the large requirement for the data required and onerous labeling by existing methods to obtain significant accuracy. Lastly, a paper [7] tackles the challenging task of aligning multiple cameras into a united coordinate system. In particular, the authors address the task of obtaining a cross-view between UAV deployed cameras and ground ones, creating an air-to-ground correspondence. The proposed solution is composed of methods that can create elaborate spatiotemporal feature maps and their cross-view space matching. This capability allows multiple cameras in a large-scale environment to be aligned into one coordination system with UAV auxiliary linkage. Therefore, such a development represents the relevant potential to enhance fields such as security surveillance, automatic control, and intelligent transportation.

Funding: This research received no external funding.

Acknowledgments: The guest editors of this Special Issue would like to thank all the authors for contributing to this volume and sharing their scientific results and experiences. We would also like to thank the journal editorial board and reviewers for conducting the review process.

Conflicts of Interest: The authors declare no conflict of interest.

References

1. Ulhaq, A.; Adams, P.; Cox, T.E.; Khan, A.; Low, T.; Paul, M. Automated Detection of Animals in Low-Resolution Airborne Thermal Imagery. *Remote Sens.* **2021**, *13*, 3276. [CrossRef]
2. Zhang, T.; Zhang, X.; Ke, X. Quad-FPN: A Novel Quad Feature Pyramid Network for SAR Ship Detection. *Remote Sens.* **2021**, *13*, 2771. [CrossRef]
3. Zheng, Y.; Sun, P.; Zhou, Z.; Xu, W.; Ren, Q. ADT-Det: Adaptive Dynamic Refined Single-Stage Transformer Detector for Arbitrary-Oriented Object Detection in Satellite Optical Imagery. *Remote Sens.* **2021**, *13*, 2623. [CrossRef]
4. Khan, A.; Asim, W.; Ulhaq, A.; Robinson, R.W. A Multiview Semantic Vegetation Index for Robust Estimation of Urban Vegetation Cover. *Remote Sens.* **2022**, *14*, 228. [CrossRef]
5. Zhang, L.; Fan, Y.; Yan, R.; Shao, Y.; Wang, G.; Wu, J. Fine-Grained Tidal Flat Waterbody Extraction Method (FYOLOv3) for High-Resolution Remote Sensing Images. *Remote Sens.* **2021**, *13*, 2594. [CrossRef]
6. Shukla, S.; Tiddeman, B.; Miles, H.C. A Wide Area Multiview Static Crowd Estimation System Using UAV and 3D Training Simulator. *Remote Sens.* **2021**, *13*, 2780. [CrossRef]
7. Li, J.; Xie, Y.; Li, C.; Dai, Y.; Ma, J.; Dong, Z.; Yang, T. UAV-Assisted Wide Area Multi-Camera Space Alignment Based on Spatiotemporal Feature Map. *Remote Sens.* **2021**, *13*, 1117. [CrossRef]

Article

Automated Detection of Animals in Low-Resolution Airborne Thermal Imagery

Anwaar Ulhaq [1], Peter Adams [2], Tarnya E. Cox [3], Asim Khan [1,4], Tom Low [5] and Manoranjan Paul [1]

[1] School of Computing, Mathematics and Engineering, Charles Sturt University,
     Port Macquarie, NSW 2444, Australia; aulhaq@csu.edu.au (A.U.); mpaul@csu.edu.au (M.P.)
[2] Department of Primary Industries and Regional Development, South Perth, WA 6151, Australia;
     peter.adams@dpird.wa.gov.au
[3] Department of Primary Industries, Orange, NSW 2800, Australia; tarnya.cox@dpi.nsw.gov.au
[4] The Institute for Sustainable Industries and Liveable Cities (ISILC), Victoria University,
     Melbourne, VIC 8001, Australia
[5] Tomcat Technologies, Orange, NSW 2800, Australia; tom@kargow.com
*   Correspondence: asim.khan@vu.edu.au

Abstract: Detecting animals to estimate abundance can be difficult, particularly when the habitat is dense or the target animals are fossorial. The recent surge in the use of thermal imagers in ecology and their use in animal detections can increase the accuracy of population estimates and improve the subsequent implementation of management programs. However, the use of thermal imagers results in many hours of captured flight videos which require manual review for confirmation of species detection and identification. Therefore, the perceived cost and efficiency trade-off often restricts the use of these systems. Additionally, for many off-the-shelf systems, the exported imagery can be quite low resolution (<9 Hz), increasing the difficulty of using automated detections algorithms to streamline the review process. This paper presents an animal species detection system that utilises the cost-effectiveness of these lower resolution thermal imagers while harnessing the power of transfer learning and an enhanced small object detection algorithm. We have proposed a distant object detection algorithm named Distant-YOLO (D-YOLO) that utilises YOLO (You Only Look Once) and improves its training and structure for the automated detection of target objects in thermal imagery. We trained our system on thermal imaging data of rabbits, their active warrens, feral pigs, and kangaroos collected by thermal imaging researchers in New South Wales and Western Australia. This work will enhance the visual analysis of animal species while performing well on low, medium and high-resolution thermal imagery.

Keywords: invasive species; thermal imaging; habitat identification; deep learning; drone

Citation: Ulhaq, A.; Adams, P.; Cox, T.E.; Khan, A.; Low, T.; Paul, M. Automated Detection of Animals in Low-Resolution Airborne Thermal Imagery. *Remote Sens.* **2021**, 13, 3276. https://doi.org/10.3390/rs13163276

Academic Editor: Maria Laura Carranza

Received: 8 June 2021
Accepted: 7 August 2021
Published: 19 August 2021

Publisher's Note: MDPI stays neutral with regard to jurisdictional claims in published maps and institutional affiliations.

1. Introduction

Recent advances in remotely piloted aircraft (RPA; a.k.a. drones, unmanned aerial vehicles) and imaging technologies have enabled a marked increase in non-invasive monitoring of animals in recent years [1–4]. The addition of thermal imaging technology offers an opportunity to not only improve the detection of target species, but, in the case of fossorial animals, their habitats as well [5–7]. However, manual detection of animals, habitat identification, and estimation of population size are cumbersome as they require frame-by-frame analysis of hours of video data. Some automated approaches have been proposed recently [8–13]. However, they often lack usability due to low accuracy, ineffectiveness against occlusion, visible spectrum limitations, and low detection speed. Thus there is a need for an intelligent, fully automated detection system.

Two main factors affect the success of automated approaches: target animal size and thermal image quality. Large mammals ($\geq$350 kg) are typically obvious with strong thermal signatures and many pixels per animal (Figure 1a). Medium-bodied mammals

(15–350 kg) can also be readily identified in an automated process, provided image quality is good, and the signature is not obscured by vegetation (Figure 1b). For smaller mammals, (≤15 kg) automated identification can be difficult even with high-quality thermal imagery in ideal conditions. Thermal signatures are often weaker, and there are fewer pixels per animal (Figure 1c). As object size becomes very small, even manual identification and tagging of correct thermal signatures is problematic. As lower altitude flights often disturb animals, high altitude flights are preferred among the research community. This poses further detection quality challenges.

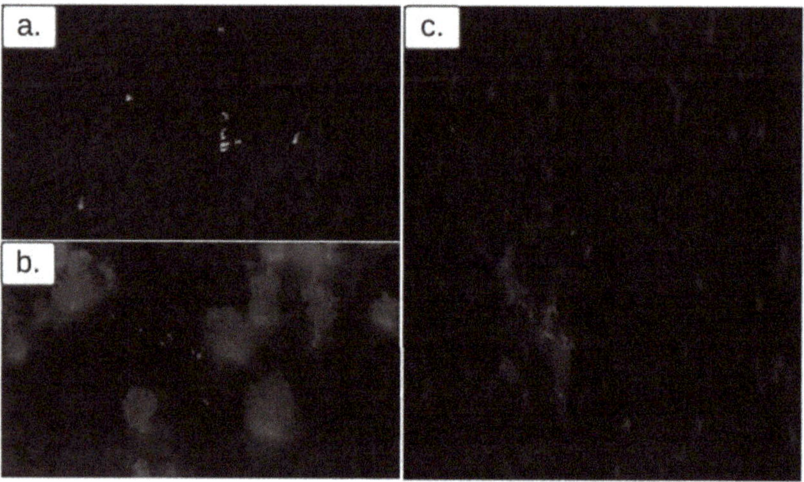

Figure 1. Larger animals ((**a**) cattle, (**b**) goats) have stronger thermal imagers and greater pixels per animal than much smaller animals ((**c**) rabbits—two individuals top left of the image). (Image (**a**,**b**) taken from footage collected on the VayuHD. Image (**c**) taken from footage collected on the Jenoptik VarioCamHD).

Deep learning has revolutionised object detection, and various deep object detection approaches exist in the literature. Some of the notable techniques include Region-based Convolutional Neural Networks (RCNN) [14], Fast-RCNN [15], Faster-RCNN [16], Mask-RCNN [17], Feature Pyramid Network (FPN) [18], Single-shot multibox Detector (SSD) [19], and You Only Look Once (YOLO) [20]. The RCNN family of detectors comprises two-stage detectors based on the concept of region proposals requiring considerable processing time and unsuitable for fast and real-time object detection. SSD [19] and YOLO [20] are one stage or one-shot detectors. SSD is very slow for detection tasks due to the sliding window approach, while YOLO outperforms these in terms of accuracy and processing time approaches. As YOLO initially is trained on the MS COCO dataset [21], its performance suffers if objects are tiny and the receptive field is limited. YOLOv3 [22] uses DarkNet-53 for feature extraction and introduces the Feature Pyramid to detect small objects at different scales. FPN predicts small-scale objects in the shallower layers with low semantic information, which might not be sufficient to classify small objects.

Our work is related to YOLO [20] and its improved versions [22,23]. Some recent work on small object detection from a distance is related to our work. An improved version of YOLO for UAV called UAV-YOLO [24] tried to improve small object detection through YOLO. It included a few more convolution layers and shortcut connections to improve the model. However, the basic limitations of subsampling remain unaddressed. In this work, we addressed the major weakness of convolution operation and aggressive subsampling and proposed a better YOLO; we called it Distant-YOLO (D-YOLO) as we detect animals from a distance.

Due to striding and pooling, the small-scale objects disappear in the deep convolution layers. Therefore, the removal of pooling and striding can improve the existing YOLO scheme to detect smaller objects. Meanwhile, YOLOv4 [23] presents new findings. However, its scope is to increase the overall speed and accuracy of the MS COCO dataset using a different bag of features and bot to increase small object detection in thermal imaging. In this work, we address the above weaknesses by introducing the proposed D-YOLO for small object detection. It enables us to propose an animal detection system with improved accuracy on imagery captured from consumer-level thermal cameras mounted on an aerial platform.

We claim the following contributions in this paper:

- We introduce animal detection from a high altitude with improved accuracy and speed using a deep learning-based object detection approach.
- We improve traditional YOLO by considering model training and structure optimisation to detect smaller and more distant objects.
- We validate our process on an extensive thermal video dataset collected by thermal imagery researchers. This dataset was very challenging as it included low resolution imagery of small animals like rabbits, and imagery of animals that, under certain conditions, can have similar thermal signatures, such as pigs and kangaroos.

2. Materials and Methodology

This section will present our data collection, data pre-processing, the proposed system architecture, and methodology. The details about each step are as follows:

2.1. Data Collection

Target species: We selected three target species for this work: the European rabbit (*Oryctolagus cuniculus*), feral pigs (*Sus scrofa*), and kangaroos (*Macropodidae*). Rabbits and pigs were selected due to the large datasets of existing thermal imagery available for use. Kangaroos were chosen as they are found in almost all habitats in Australia and are regularly captured on thermal imaging surveys for other species. To perform this study, we first established an image database. The imagery in this database was collected by the Department of Primary Industry, New South Wales (NSW), and the Department of Primary Industries and Regional Development, Western Australia.

In the proposed work, we used the deep neural network-based object detection method for animal detection in thermal imaging data.

Thermal imager types and specifications: Thermal imagery was collected via several platforms with a range of thermal imagers (Table 1). The imagers used range in price and quality of exported imagery (please see Cox et al. [7] for a discussion on the effect of these specifications on image output).These imagers were not selected to collect imagery, rather, these are the imagers that the imagery used in this research was collected with.

Table 1. The types and specifications of the thermal imagers that collected the footage that was used for this study.

Imager	Platform	View (Hz)	Export (Hz)	Sensor (w × h) (mm)	Lens (mm)	Pixel Pitch	Target Animal Species
FLIR Zenmuse	DJI Innspire-1 RPA	30	9	12.38 × 9.68	640 × 512	17μ	Rabbit and rabbit warren
Janoptik Vario CAM HD	DJI S1000+ RPA/Ground based survey	30	30	17.4 × 9.68	1024 × 800	17μ	Rabbit
Sierra Olympic Vayu HD	DJI M600 RPA	60	>30	24 × 14.5	1920 × 1200	12μ	Rabbit, rabbit warren, pigs and kangaroo
FLIR Zenmuse XT 640	DJI Matrice 210 RPA	9	9	12.38 × 9.68	640 × 512	17μ	Pigs and Kangaroo

2.2. Data Pre-Processing

From the thermal footage obtained, we extracted frames to prepare the training dataset. As the video frame rate from the Vayu (used for the training dataset) is 60 fps, we had a huge number of extracted frames. However, most frames have no evidence of any animals; therefore, we used only those frames that had confirmed the presence of targeted animals while discarding the rest of the frames in feeding our training model for robust results.

For supervised training, we manually labelled the dataset. We used the python-based library open-source annotation tool "Labelme", a graphical image annotation tool inspired by MIT, Computer Science and Artificial Intelligence Laboratory [25]. We also observed that target objects were very small in some of the frames collected from a high altitude (67 m). Similarly, some of the targets were obscure, and even manual classification of their thermal signatures was challenging. We had to magnify such frames/images to label them accurately. Some sample shots of the manual annotation of our thermal dataset are shown in Figure 2, whereas Table 2 illustrates the dataset details.

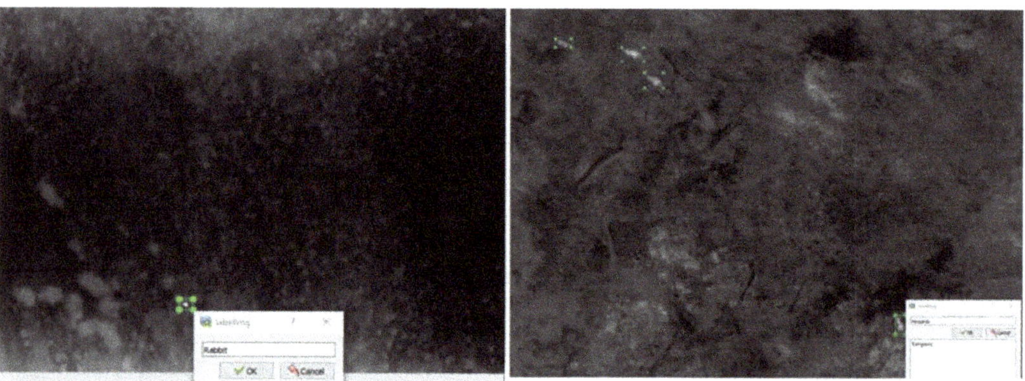

Figure 2. An example of data annotation/labelling performed for different animal species in our dataset.

Table 2. Dataset used for training purpose.

Class Name	Labelled	Total Images
Rabbit	Rabbit	1246
Kangaroos	Kangaroo	4211
Pigs	Pig	6000

2.3. Data Annotation, Model Training and Detection

Our footage library was extensive; thus, we divided it into three datasets: training dataset, evaluation dataset, and testing dataset. First, we annotated target animals of interest in our training dataset using a Python-based annotation tool. We then trained our proposed D-YOLO model on the training dataset. A detailed description of D-YOLO is provided below.

During data collection, we took both far and nearer footage of animals using different camera zoom. Therefore, to improve the performance of YOLOv3 for small object detection, we divided our dataset into two categories named "zoom-out" and "zoom-in" groups by taking the distance and receptive field into consideration, as shown in Figure 3. We also used data augmentation to balance their sizes. K-means [26] was then used to cluster different numbers of anchor boxes to find the optimised number and size for better results. Finally, the model was retrained using the "zoom-out" category data.

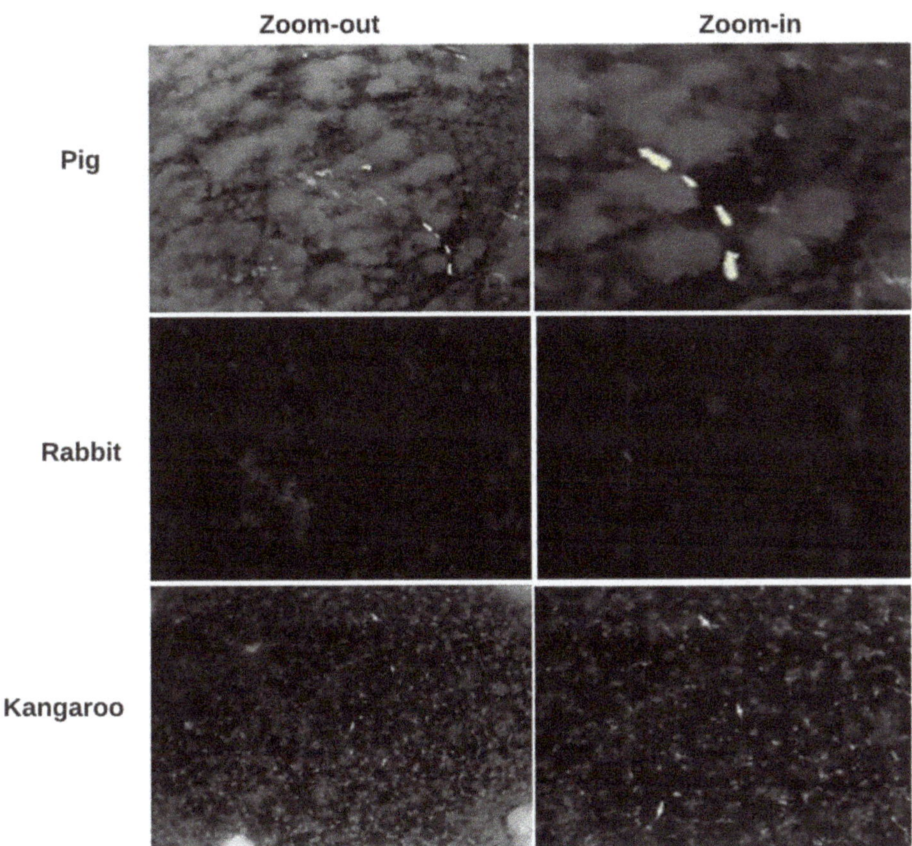

Figure 3. Dataset sample of zoom-in and zoom-out.

A brief introduction to YOLO: YOLOv3 is a more established one-shot detector that is an incremental model of the former YOLO [27], and YOLO9000 [20]. The YOLOv3 backbone known as DarkNet-53 includes 53 convolution layers and Resnet [28] short cut connections. The prediction stage uses FPN that uses three scale feature maps, where small feature maps provide semantic information and large feature maps provide finer-grained information. Darknet (conv2D BN Leaky, short as DBL) comprises one convolution layer, one batch normalisation layer, and one leaky relu layer displayed as DBL. YOLOv3 uses independent logistic classifiers rather than softmax with binary cross-entropy loss for the class predictions in the training stage. FPN uses three detection scales with different receptive fields, where the 32-fold down-sampling is suitable for large objects, the 16-fold for middle-sized objects, and the 8-fold for small size objects.

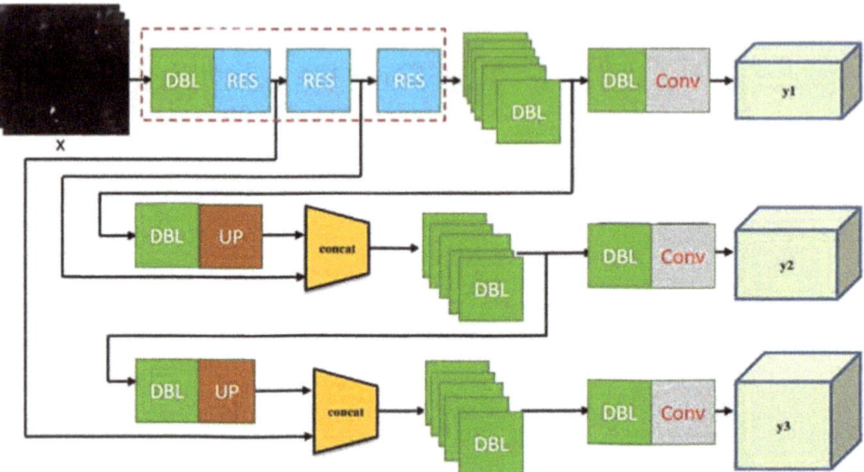

Figure 4. YOLOv3 architecture with the input image and three types of feature map as output. The basic element of YOLOv3 is called Darknet conv2D BN Leaky (DBL), which is composed of one convolution layer, one batch normalization layer, and one leaky relu layer. Other important components of the structure are explained in the text.

An architectural diagram of YOLOv3 is shown in Figure 4. It takes an input image of size 416 × 416 pixels and calculates three types of feature map (13 × 13 × 3, 26 × 26 × 3, and 52 × 52 × 3) bounding boxes as output. Darknet (conv2D BN Leaky, shortened to DBL) comprises one convolution layer, one batch normalisation layer, and one leaky relu layer displayed as DBL. It also includes ResUnit that includes two "DBL" structures followed by one "add" layer. It leads to the residual-like unit, "ResBlock". "ResBlock" is the module element of Darknet 53.

The proposed D-YOLO Scheme:

One of the problems with traditional CNN networks is their inability to handle low resolution and receptive field at both pooling and striding may cause loss of small targets. The semantic information about the small objects will vanish or weaken with a decreased spatial resolution of feature maps in subsequent layers. Low semantic information may not be enough to recognise the small object category in thermal images.

A region of the input on which a pixel value in the output depends is called the receptive field. CNN's pooling (progressively reducing resolution and removing sub-sampling) can help, but it reduces the receptive field. On the other hand, dilated convolutions [29] can increase the explanation of the output feature maps without harming the receptive field of individual neurons. Dilated convolution is also called "convolution with a dilated filter", as it is a similar filter used for wavelet transformation. This concept is explained in Figure 5.

Let $F : Z^2 \rightarrow R$ be a discrete function, $\Phi_n = \lceil -n, n \rceil^2$ and let $f = \Phi_n \rightarrow R$ be another discrete function; the convolution operator $*$ can be defined as :

$$(F * f)(x) = \sum_{s+t=x} F(s)f(t) \tag{1}$$

Let us define d as a dilation factor and let $*_d$ be defined as:

$$(F *_d f)(x) = \sum_{s+dt=x} F(s)f(t) \tag{2}$$

where $*_d$ is a d-dilated convolution, the traditional CNN convolution is simply the 1-dilated convolution. Dilated convolution supports an exponential expansion of the receptive field without loss of resolution. Figure 5 illustrated the outcome of dilated convolution. F1,

F2 are the larger grid showing original discrete functions and f1, f2 are the green colour discrete filters. Figure 5A on the left shows output generated from convolving F1 by a 1-dilated convolution f1; where F is the larger grid and f is the green colour filter. Each element in this representation has a receptive field of 3 × 3. Figure 5B on the right shows the output generated from F2 convolved with a 2-dilated convolution f2; Each element in this representation has a receptive field of 7 × 7.

Therefore, to increase the receptive field of YOLO to handle small objects, we integrated dilated convolutions in its architecture. For this purpose, we replaced the DDL block with a DDDL block that uses dilated convolution followed by batch normalisation and leaky Relu. Likewise, RES block is replaced with DRN (Dilated Residual Network) [30]. Similarly, for multiscale spatial pooling, we use different dilation rates and replace upsampling with dilation filtering. Finally, semantic information from three scales is concatenated to detect objects and their categories. The proposed D-YOLO architecture is shown in Figure 6.

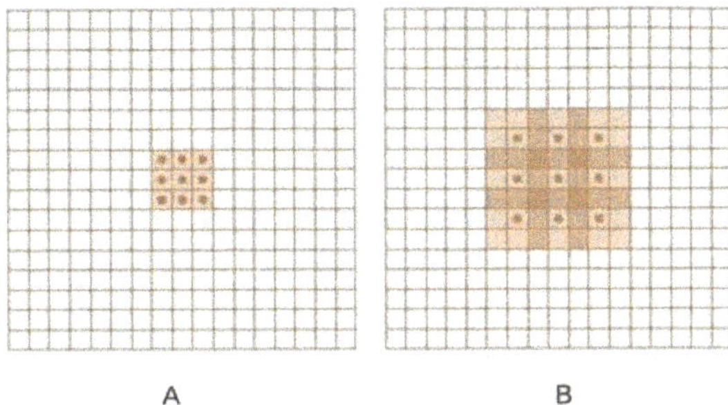

A   B

Figure 5. Dilated Convolutions: The figure (**A**) on the left shows the output generated from convolving F1 by a 1-dilated convolution f1; F1, F2 are the larger grid showing original discrete functions and f1, f2 are the green colour discrete filters. Each element in this representation has a receptive field of 3 × 3. Figure (**B**) on the right shows the output generated from F2 convolved with a 2-dilated convolution f2; Each element in this representation has a receptive field of 7 × 7.

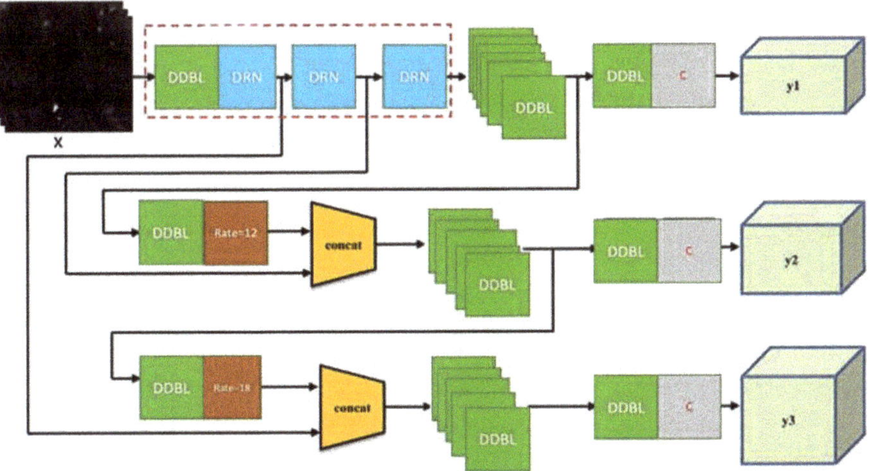

Figure 6. D-YOLO architecture with input image size 416 × 416 pixels and 3 types of feature map (13 × 13 × 3, 26 × 26 × 3, and 52 × 52 × 3) as output; DDBL stands for Darknet dilated conv2D BN Leaky, composed of one convolution layer, one batch normalisation layer, and one leaky relu layer.; DRN (Dilated Residual Network) provides residual-like connection with dilated convolutions. Similarly, for multiscale spatial pooling, we use different dilation rates and replace upsampling with dilation filtering.

2.4. Geo-Tagging and Visualizing of Detected Targets

Finally, geo-tagging of detected animals is done by embedding a Google maps API platform on the acquired flight GPS data for locating and visualising targets in processing real-time. It provides precise tracking of target locations and visualisation of their movement within the surrounding. Such information is key to monitor animal movement patterns and gain valuable insights about their activities. Figure 7 illustrates the process of geo-tagging of detected animals from aerial data and also provides visualisation of their movements during the time of flight.

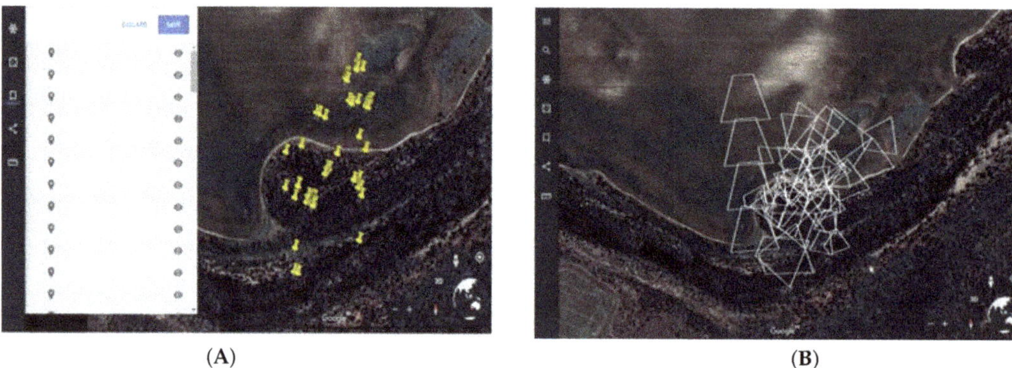

(A) (B)

Figure 7. (**A**) The geo-tagging of detected animals from drone data that points out their detection location and (**B**) visualisation of animal movements that display the area of their activity.

2.5. Experiments and Results

The majority of classifiers assume that output labels are mutually exclusive. If the output consists of mutually exclusive object classes, this is true. As a result, YOLO uses a softmax function to transform scores into one-to-one probability. At this point, the total output can be larger than one. The algorithm substitutes the softmax algorithm with independent logistic classifiers to assess the likelihood that an input belongs to a certain label. The algorithm calculates the classification loss for each tag using binary cross-entropy loss rather than mean square error. Omitting the softmax function also reduces processing complexity.

The algorithm uses logistic regression to estimate an objectness score for each bounding box. The matching objectness score should be one of the bounding boxes prior (anchor) overlapping a ground truth object more than others. Only one boundary box prior is linked with each ground truth item. There is no classification or localisation loss if a bounding box prior is not assigned; nonetheless, there is a confidence loss on objectness. To compute the loss, we utilise tx and ty (rather than bx and by).

$$b_x = \sigma(t_x) + c_x$$
$$b_y = \sigma(t_y) + c_x$$
$$b_w = \rho_w e^{t_w}$$
$$b_h = \rho_h e^{t_h}$$

Precision, recall, accuracy, and the f1-score are some of the ways used to examine the performance of neural networks. The precision tells us about the correct predictions made out of false-positives, while recall tells us about the correct predictions made out of false negatives. The accuracy is the number of correct predictions out of both false positives and false negatives. All the performance metrics for our trained model have been determined using the formulas listed in Equations (3)–(6).

$$Precision = \frac{TP}{TP + FP} \tag{3}$$

$$Recall = \frac{TP}{TP + FN} \tag{4}$$

$$Accuracy = \frac{TP + TN}{TP + TN + FN + FP} \tag{5}$$

$$F1 - Score = 2 * \frac{precision * recall}{precision + recall} \tag{6}$$

where true positives are *TP*, true negatives are *TN*, false positives are *FP*, and false negatives are *FN*. The *TP* and *TN* are the right predictions, whereas the *FP* and *FN* are our model's wrong predictions.

We carried out the training process for our deep model experiments both on Windows and Ubuntu operating systems. We used the deep learning framework PyTorch and related Python libraries for system training and testing. Training and testing were performed on both windows and ubuntu operating systems workstations. They had an Intel ninth gen i9 CPU, i.e., 9900 k, 64 GB RAM and Nvidia dual RTX 2080 Ti 11 GB VRAM GPUs. Table 3. shows the system specifications.

Table 3. System Specifications for Training/Testing.

System Hardware/Software (Operating System)	Specifications
RAM	64 GB RAM
CPU	Intel 9th Gen i9 9900K
GPU(s)	2x NVIDIA RTX 2080 Ti 11 GB VRAM
Operating System	Windows 10 Professional and Ubuntu 18.04

The experimental dataset was divided into training and validation as 85% and 15%, respectively, as shown in Table 4 to get the optimised results and overcome the issue of over-fitting.

Table 4. Data Split for Testing/Training & Accuracy Obtained.

Dataset (Train/Test) Split in %	Accuracy (%)				
	10 Epochs	20 Epochs	30 Epochs	40 Epochs	50 Epochs
85–15	92.31	95.84	96.86	97.39	98.38

We first tried to establish the baseline by training a YOLOv3 based detection; for this purpose, we used the size of input frames as an integer multiple of 32 (416 × 416), with five steps for downsampling operation leading to the largest stride size of 32. As this version used multi-scale analysis, y1, y2, and y3 lead to three different feature maps. Information for the detection of final bounding boxes comes from the combination of all three scales.

We fine-tuned a pre-trained YOLOv3 model for training, with a mini-batch size of 32, 10,500 batches, subdivisions of 15 on 1 GPU, a momentum of 0.8, and a weight decay of 0.0004. We adopted the multistep learning rate with a base learning rate of 0.0001 and the learning rate scales of [0.1, 0.1].

We then designed the proposed D-YOLO algorithm by replacing convolutions with dilated versions. For this purpose, we used the size of input frames as an integer multiple of 32 (416 × 416), without a downsampling operation, and introduced dilation rates of 6, 12, and 18 at different levels. The rest of the design remains the same. Information for the detection of final bounding boxes comes from the combination of all three scales. However, the original model size remains the same as of YOLOv3. We used similar training specifications for our baseline model.

The average of numerous intersections over union (IoU) is referred to as the average precision (AP) (the minimum IoU to consider a positive match). For example, AP@[.5:.95] represents the average AP for IoU with a step size of 0.05 from 0.5 to 0.95. In our experiments, the mAP0.5 is 0.871, as shown in Figure 8a. We achieved an average accuracy of 98.33% for the D-YOLO during the testing phase, compared to 92.33% accuracy for the baseline YOLO model. For Pig class accuracy = 97.34%, recall = 96.89%, precision = 96.37%, and f1-score = 96.35%. Kangaroo class accuracy = 99.48%, recall = 96.96%, precision = 97.30%, and f1-score = 98.60%. Rabbit class accuracy = 98.17%, recall = 96.70%, precision = 96.48%, and f1-score = 97.48%. Figure 8b visualizes the above results. Hence, kangaroo signatures are bigger and differentiable, and therefore achieved better accuracy for this class. For warren detection, we achieved (accuracy = 93.34%, recall = 96.89%, precision = 96.37%, and f1-score = 96.3%).

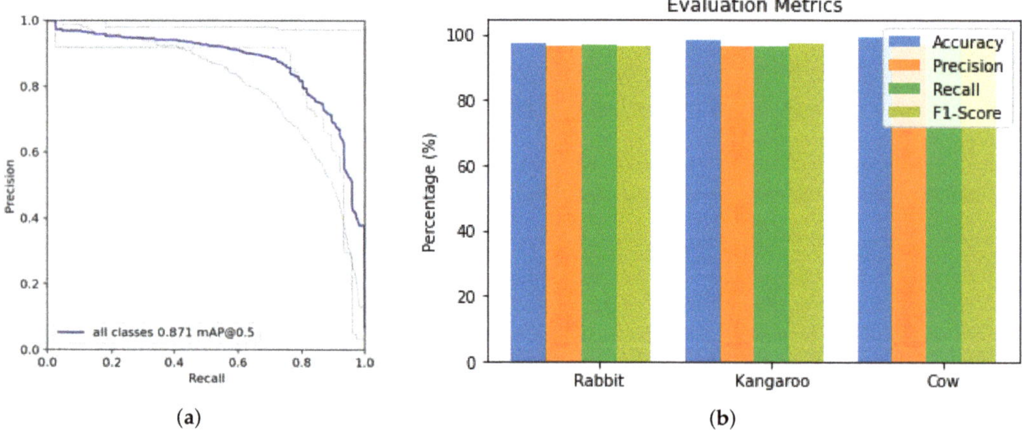

Figure 8. (**a**) Mean Average Precision (mAP) for all the classes. (**b**) System performance metrics for each class of animal.

Training and test accuracy was calculated for our training and validation set. Figure 9a,b displays our training and validation loss for each epoch. These graphs were generated for a data split of 85–15%. The accuracy graph visually shows that both training and testing accuracy increases gradually and then converges on a specific point. It also shows that after 40 epochs, the accuracy reduction reduces as the validation accuracy appears to be equivalent to training accuracy. Similarly, the right graph shows how the loss decreases gradually as the model learns on a given dataset. The loss of validation data becomes stable after 43 epochs and thus tends towards a specific value.

We tested our approach on the data that was not part of our training or validation set. We first detected all bounding boxes and used them for counting the number of detected animals. Then, to remove double counting, we sustained our count until the 10th frame. This value was found empirically based on manual inspection of frames and detected animals. Finally, we counted ground truth detections and compared them with the automated population count of animals for verification purposes. This process also verified our detection results accuracy.

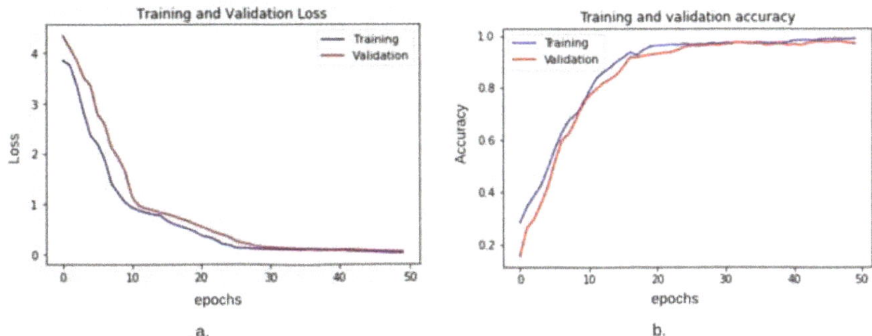

a.

b.

Figure 9. (**a**). Training and validation loss plot and (**b**). Training and validation accuracy plot. Both the plots show consistency in decreasing loss and increasing accuracy on the given dataset. Training is shown in blue colour, and validation is shown in red colour.

Some of the sample detection results are shown in Figure 10. Detected labels and their sizes are intentionally made small to show small bounding boxes.

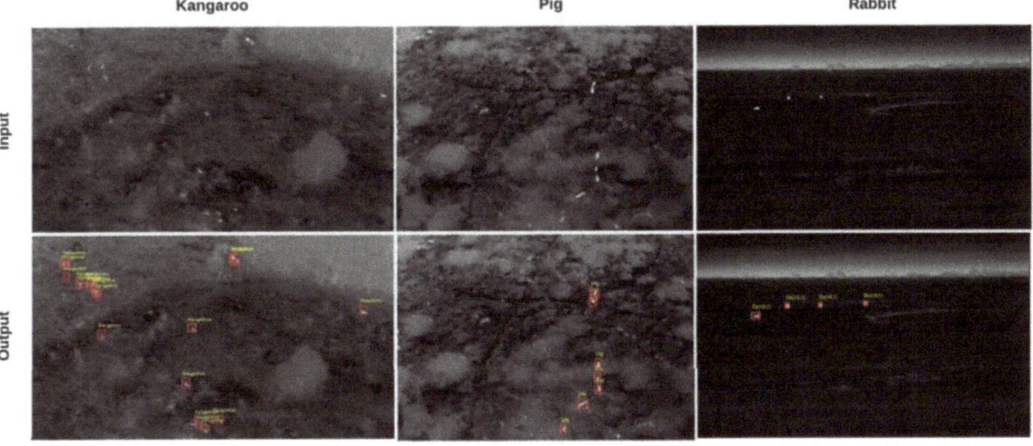

Figure 10. Sample Results: the first row includes input images, while the second row shows respective output images. Both bounding boxes and labels are shown.

Similarly, we also trained our model for automated identification of habitats specifically rabbit warrens. Where rabbit numbers are high, warren entrances tend to be visible on thermal imagery, Cox et al. [7]. Figure 11 and so could be included in the automated detection with confidence.

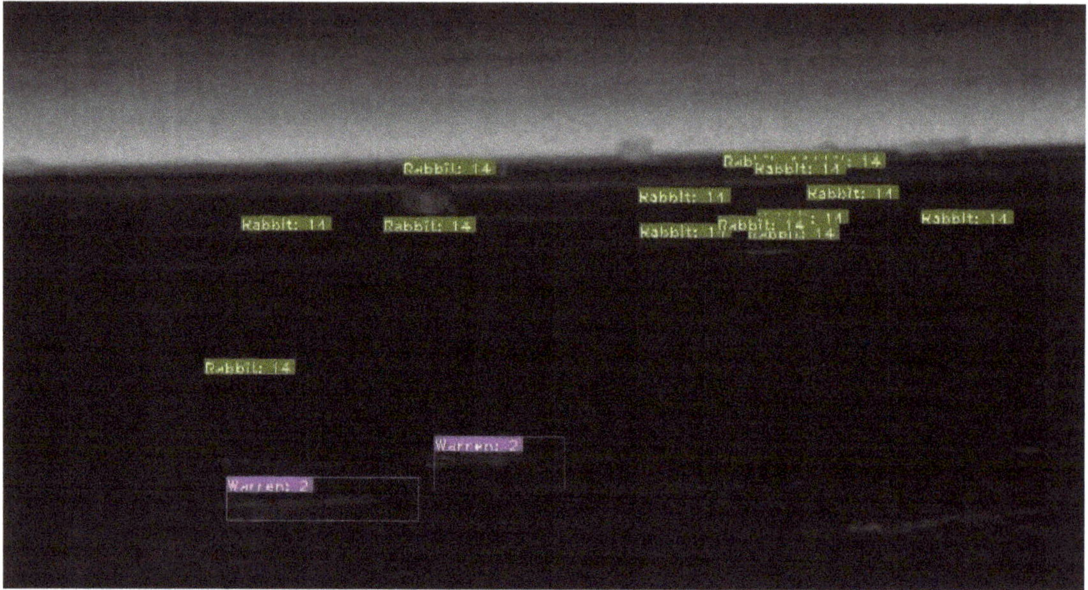

Figure 11. Identification of rabbits and their warren is shown with their respective labels found by our model. Yellow colour labels belong to rabbits, while purple colour labels are their warrens.

3. Dicussion

Our system enabled detection of small animals in low-resolution video sequences from thermal imagery. As thermal imagery is used more widely in wildlife management programs, the ability to accurately identify animals within the footage in timely manner will only become more important. Our work enhances the existing object detection algorithm, YOLO, to work with low resolution thermal imagery on a select number of animals. Future work will see the current model extended to include several other species of animal to make this system more broadly applicable.

We successfully detected all animals with D-YOLO that were detected manually. However, it is not yet known whether D-YOLO is better at detecting animals in thermal imagery than manual detection. Manual detection relies on observers, and observers can be subject to biases and other factors, including fatigue, interest, skill level, training, eyesight etc. These factors are essentially removed in automated detection, so automated detection may be more effective at identifying target species than manual review. Further investigation of how D-YOLO performs compared to manual detection is required.

The comparison of automated and manual review has flow-on effects for how automated detection algorithms are used in wildlife research programs. For pest animal management, the cost of missing or underestimating the number of animals in the landscape can be more than relying on manual detection. In these cases, there needs to be high confidence in the accuracy and precision of the algorithm compared to manual review. However, in conservation scenarios, this level of accuracy and precision may be less important. Therefore, not detecting every animal available does not have a negative cost impact on the land manager.

While we successfully developed this algorithm to identify these three species correctly, an aspect requiring further development is removing double counts. Although we didn't report on this here, the same animal was counted twice in some instances as animals were identified. Future work will focus on developing robust strategies to manage this problem so that accurate counts of species can also be provided. Nevertheless, it is an important element of automation and will provide wildlife researchers and managers with a functional and valuable tool for reviewing thermal imaging footage.

4. Conclusions

This paper proposed a robust detection system to identify animals, and in the case of rabbits, their habitat, from aerial thermal imaging data. Our dataset had several challenges as the size of target animals was cryptic and small, but the resolution of our cameras was also low. This project aimed to develop a robust system for the identification of animals using consumer-level cameras. For this purpose, we introduced the distant object detection algorithm named D-YOLO (Distant-You Only Look Once) [20] for remote detection of small targets. Our system, trained on the massive data collected from New South Wales and Western Australia, can detect animals (rabbits, kangaroos, and pigs) with a probablity comparable to that of manual detection. This work will facilitate wildlife researchers to monitor the activities of animals across the landscape.

Author Contributions: All authors have made significant contributions to this research. Conceptualization, A.U. and A.K.; methodology, A.U., A.K. and T.L.; software, A.K. and T.L.; validation, A.U. and A.K.; formal analysis, A.K. and T.L.; investigation, A.U., A.K., M.P., P.A., T.E.C. and T.L.; resources, P.A. and T.E.C., data curation, A.K. and T.L.; writing—original draft preparation, A.U. and A.K.; writing—review and editing, A.U., A.K., M.P., P.A. and T.E.C.; visualization, A.K. and T.L.; supervision, A.U., A.K. and M.P.; project administration, A.U., A.K., M.P., P.A., T.E.C. and T.L.; funding acquisition, P.A. and T.E.C. All authors have read and agreed to the published version of the manuscript .

Funding: This research work was funded through the Australian Commonwealth Government's Control tools and technologies for established pest animals and weeds competitive grants program 2017 and was completed with animal ethics approval (Orange AEC-ORA18/21/021).

Institutional Review Board Statement: Not applicable.

Informed Consent Statement: Not applicable.

Data Availability Statement: The data used to support the findings of this study are available subject to approval from the relevant departments through the corresponding author upon request.

Conflicts of Interest: The authors declare no conflict of interest.

References

1. Van Hespen, R.; Hauser, C.E.; Benshemesh, J.; Rumpff, L.; Monfort, J.J.L. Designing a camera trap monitoring program to measure efficacy of invasive predator management. *Wildl. Res.* **2019**, *46*, 154–164. [CrossRef]
2. Jepsen, E.M.; Ganswindt, A.; Ngcamphalala, C.A.; Bourne, A.R.; Ridley, A.R.; McKechnie, A.E. Non-invasive monitoring of physiological stress in an afrotropical arid-zone passerine bird, the southern pied babbler. *Gen. Comp. Endocrinol.* **2019**, *276*, 60–68. [CrossRef] [PubMed]
3. Georgieva, M.; Georgiev, G.; Mirchev, P.; Filipova, E. Monitoring on appearance and spread of harmful invasive pathogens and pests in Belasitsa Mountain. In Proceedings of the X International Agriculture Symposium, Agrosym 2019, Jahorina, Bosnia and Herzegovina, 3–6 October 2019; Faculty of Agriculture, University of East Sarajevo: Lukavica, Bosnia and Herzegovina, 2019; pp. 1887–1892.
4. Burke, C.; Rashman, M.; Wich, S.; Symons, A.; Theron, C.; Longmore, S. Optimizing observing strategies for monitoring animals using drone-mounted thermal infrared cameras. *Int. J. Remote Sens.* **2019**, *40*, 439–467. [CrossRef]
5. Witczuk, J.; Pagacz, S.; Zmarz, A.; Cypel, M. Exploring the feasibility of unmanned aerial vehicles and thermal imaging for ungulate surveys in forests-preliminary results. *Int. J. Remote Sens.* **2018**, *39*, 5504–5521. [CrossRef]
6. Colomina, I.; Molina, P. Unmanned aerial systems for photogrammetry and remote sensing: A review. *ISPRS J. Photogramm. Remote Sens.* **2014**, *92*, 79–97. [CrossRef]
7. Cox, T.E.; Matthews, R.; Halverson, G.; Morris, S. Hot stuff in the bushes: Thermal imagers and the detection of burrows in vegetated sites. *Ecol. Evol.* **2021**, *11*, 6406–6414. [CrossRef] [PubMed]
8. Karp, D. Detecting small and cryptic animals by combining thermography and a wildlife detection dog. *Sci. Rep.* **2020**, *10*, 5220. [CrossRef] [PubMed]
9. Berg, A.; Johnander, J.; Durand de Gevigney, F.; Ahlberg, J.; Felsberg, M. Semi-automatic annotation of objects in visual-thermal video. In Proceedings of the IEEE International Conference on Computer Vision Workshops, Seoul, Korea, 27–28 October 2019.

10. Kellenberger, B.; Marcos, D.; Lobry, S.; Tuia, D. Half a percent of labels is enough: Efficient animal detection in UAV imagery using deep cnns and active learning. *IEEE Trans. Geosci. Remote Sens.* **2019**, *57*, 9524–9533. [CrossRef]
11. Meena, D.; Agilandeeswari, L. Invariant Features-Based Fuzzy Inference System for Animal Detection and Recognition Using Thermal Images. *Int. J. Fuzzy Syst.* **2020**, *22*, 1868–1879. [CrossRef]
12. Shepley, A.J.; Falzon, G.; Meek, P.; Kwan, P. Location Invariant Animal Recognition Using Mixed Source Datasets and Deep Learning. *bioRxiv* **2020**. [CrossRef]
13. Corcoran, E.; Denman, S.; Hanger, J.; Wilson, B.; Hamilton, G. Automated detection of koalas using low-level aerial surveillance and machine learning. *Sci. Rep.* **2019**, *9*, 3208. [CrossRef] [PubMed]
14. Girshick, R.; Donahue, J.; Darrell, T.; Malik, J. Rich feature hierarchies for accurate object detection and semantic segmentation. In Proceedings of the IEEE Conference on Computer Vision and Pattern Recognition, Columbus, OH, USA, 23–28 June 2014; pp. 580–587.
15. Girshick, R. Fast r-cnn. In Proceedings of the IEEE International Conference on Computer Vision, Santiago, Chile, 7–13 December 2015; pp. 1440–1448.
16. Ren, S.; He, K.; Girshick, R.; Sun, J. Faster r-cnn: Towards real-time object detection with region proposal networks. *Adv. Neural Inf. Process. Syst.* **2015**, *28*, 91–99. [CrossRef] [PubMed]
17. He, K.; Gkioxari, G.; Dollár, P.; Girshick, R. Mask r-cnn. In Proceedings of the IEEE International Conference on Computer Vision, Venice, Italy, 22–29 October 2017; pp. 2961–2969.
18. Lin, T.Y.; Dollár, P.; Girshick, R.; He, K.; Hariharan, B.; Belongie, S. Feature pyramid networks for object detection. In Proceedings of the IEEE Conference on Computer Vision and Pattern Recognition, Honolulu, HI, USA, 21–26 July 2017; pp. 2117–2125.
19. Liu, W.; Anguelov, D.; Erhan, D.; Szegedy, C.; Reed, S.; Fu, C.Y.; Berg, A.C. Ssd: Single shot multibox detector. In *European Conference on Computer Vision*; Springer: Berlin/Heidelberg, Germany, 2016; pp. 21–37.
20. Redmon, J.; Farhadi, A. YOLO9000: Better, faster, stronger. In Proceedings of the IEEE Conference on Computer Vision and Pattern Recognition, Honolulu, HI, USA, 21–26 July 2017; pp. 7263–7271.
21. Lin, T.Y.; Maire, M.; Belongie, S.; Hays, J.; Perona, P.; Ramanan, D.; Dollár, P.; Zitnick, C.L. Microsoft coco: Common objects in context. In *European Conference on Computer Vision*; Springer: Berlin/Heidelberg, Germany, 2014; pp. 740–755.
22. Redmon, J.; Farhadi, A. Yolov3: An incremental improvement. *arXiv* **2018**, arXiv:1804.02767.
23. Bochkovskiy, A.; Wang, C.Y.; Liao, H.Y.M. YOLOv4: Optimal Speed and Accuracy of Object Detection. *arXiv* **2020**, arXiv:2004.10934.
24. Liu, M.; Wang, X.; Zhou, A.; Fu, X.; Ma, Y.; Piao, C. UAV-YOLO: Small Object Detection on Unmanned Aerial Vehicle Perspective. *Sensors* **2020**, *20*, 2238. [CrossRef]
25. LabelMe. The Open Annotation Tool. Available online: http://labelme2.csail.mit.edu/Release3.0/index.php?message=1 (accessed on 9 July 2021).
26. Guo, W.; Li, W.; Gong, W.; Cui, J. Extended Feature Pyramid Network with Adaptive Scale Training Strategy and Anchors for Object Detection in Aerial Images. *Remote Sens.* **2020**, *12*, 784. [CrossRef]
27. Redmon, J.; Divvala, S.; Girshick, R.; Farhadi, A. You only look once: Unified, real-time object detection. In Proceedings of the IEEE Conference on Computer Vision and Pattern Recognition, Las Vegas, NV, USA, 27–30 June 2016; pp. 779–788.
28. He, K.; Zhang, X.; Ren, S.; Sun, J. Deep residual learning for image recognition. In Proceedings of the IEEE Conference on Computer Vision and Pattern Recognition, Las Vegas, NV, USA, 27–30 June 2016; pp. 770–778.
29. Yu, F.; Koltun, V. Multi-scale context aggregation by dilated convolutions. *arXiv* **2015**, arXiv:1511.07122.
30. Yu, F.; Koltun, V.; Funkhouser, T. Dilated residual networks. In Proceedings of the IEEE Conference on Computer Vision and Pattern Recognition, Honolulu, HI, USA, 21–26 July 2017; pp. 472–480.

Article

Quad-FPN: A Novel Quad Feature Pyramid Network for SAR Ship Detection

Tianwen Zhang, Xiaoling Zhang * and Xiao Ke

School of Information and Communication Engineering, University of Electronic Science and Technology of China, Chengdu 611731, China; twzhang@std.uestc.edu.cn (T.Z.); xke@std.uestc.edu.cn (X.K.)
* Correspondence: xlzhang@uestc.edu.cn

Abstract: Ship detection from synthetic aperture radar (SAR) imagery is a fundamental and significant marine mission. It plays an important role in marine traffic control, marine fishery management, and marine rescue. Nevertheless, there are still some challenges hindering accuracy improvements of SAR ship detection, e.g., complex background interferences, multi-scale ship feature differences, and indistinctive small ship features. Therefore, to address these problems, a novel quad feature pyramid network (Quad-FPN) is proposed for SAR ship detection in this paper. Quad-FPN consists of four unique FPNs, i.e., a DEformable COnvolutional FPN (DE-CO-FPN), a Content-Aware Feature Reassembly FPN (CA-FR-FPN), a Path Aggregation Space Attention FPN (PA-SA-FPN), and a Balance Scale Global Attention FPN (BS-GA-FPN). To confirm the effectiveness of each FPN, extensive ablation studies are conducted. We conduct experiments on five open SAR ship detection datasets, i.e., SAR ship detection dataset (SSDD), Gaofen-SSDD, Sentinel-SSDD, SAR-Ship-Dataset, and high-resolution SAR images dataset (HRSID). Qualitative and quantitative experimental results jointly reveal Quad-FPN's optimal SAR ship detection performance compared with the other 12 competitive state-of-the-art convolutional neural network (CNN)-based SAR ship detectors. To confirm the excellent migration application capability of Quad-FPN, the actual ship detection in another two large-scene Sentinel-1 SAR images is conducted. Their satisfactory detection results indicate the practical application value of Quad-FPN in marine surveillance.

Keywords: synthetic aperture radar (SAR); ship detection; convolutional neural network (CNN); deep learning (DL); feature pyramid network (FPN); quad feature pyramid network (Quad-FPN)

Citation: Zhang, T.; Zhang, X.; Ke, X. Quad-FPN: A Novel Quad Feature Pyramid Network for SAR Ship Detection. *Remote Sens.* **2021**, *13*, 2771. https://doi.org/10.3390/rs13142771

Academic Editors: Anwaar Ulhaq and Douglas Pinto Sampaio Gomes

Received: 26 May 2021
Accepted: 2 July 2021
Published: 14 July 2021

Publisher's Note: MDPI stays neutral with regard to jurisdictional claims in published maps and institutional affiliations.

1. Introduction

Synthetic aperture radar (SAR) is an advanced active microwave sensor for the high-resolution remote sensing observation of the Earth [1]. Its all-day and all-weather working capacity makes it play an important role in marine surveillance [2]. As a fundamental marine mission, SAR ship detection is of great value in marine traffic control, fishery management, and emergent salvage at sea [3,4]. Thus, up to now, the topic of SAR ship detection has received continuous attention from an increasing number of scholars [5–15].

In earlier years, a standard solution is to design ship features by manual ways, e.g., constant false alarm rate (CFAR) [1], saliency [2], super-pixel [3], and transformation [4]. Yet, these traditional methods are always complex in algorithm, weak in migration, and cumbersome in manual design, leading to their limited migration applications. Moreover, they often use limited ship images for theoretical analysis to define ship features, but these features cannot reflect the characteristics of ships with various sizes under different backgrounds. This causes their poor multi-scale and multi-scene detection performance.

Fortunately, in recent years, with the rise of deep learning (DL) and convolutional neural networks (CNNs), current state-of-the-art DL-based/CNN-based SAR ship detectors have helped solve the above-mentioned problems, to some degree. Compared with traditional methods, CNN-based ones have significant advantages, i.e., simplicity, high-efficiency, and high-accuracy, because they can enable computational models with multiple processing layers to learn data representations with multiple-level abstractions. This can effectively improve detection accuracy. Thus, nowadays, many scholars [5–15] in the SAR ship detection community are starting to pay much attention to CNN-based methods.

For instance, based on Fast R-CNN [16], Li et al. [9] proposed a binarized normed gradient-based method to extract SAR ship-like regions. Based on Faster R-CNN [17], Lin et al. [14] designed a squeeze and excitation rank mechanism to improve detection performance. Based on you only look once (YOLO) [18], Zhang et al. [10] integrated the multi-scale mechanism, concatenation mechanism, and anchor box mechanism for small ship detection. Based on RetinaNet [19], Yang et al. [11] tried to suppress ship detections' false alarms by loss weighting means. Based on single shot multi-box detector (SSD) [20], Wang et al. [7] proposed an optimized version to enhance small ship detection while improving detection speed. Based on Cascade R-CNN [21], Wei et al. [12] designed a robust SAR ship detector named HR-SDNet for multi-level ship feature extraction.

Since the feature pyramid network (FPN) was proposed by Lin et al. [22], it has been a standard solution for multi-scale SAR ship detection. For different resolutions, incident angles, satellites, etc., SAR ships possess various sizes. FPN can detect ships with different sizes at different resolution levels based on more reasonable semantic features from backbone networks. This enables better detection performance. Thus, it has received a wide range of attention, e.g., Wei et al. [12] optimized its structure to present a high-resolution FPN for better multi-scale detection. Cui et al. [13] adopted a convolutional block attention module to improve its performance. Lin et al. [14] added a squeeze-and-excitation module at the top of FPN to activate important features. Zhao et al. [15] designed an attention receptive pyramid network to detect ships with various sizes and complex backgrounds.

However, SAR ship detection is still a challenging issue due to complex background interferences (e.g., port facilities, sea clutters, and volatile sea states), multi-scale ship feature differences, and indistinctive small ship features. Thus, this paper proposes a novel quad feature pyramid network (Quad-FPN) for SAR ship detection. Figure 1 shows Quad-FPN's structure. From Figure 1, four FPNs constitute it, i.e., a DEformable COnvolutional FPN (DE-CO-FPN), a Content-Aware Feature Reassembly FPN (CA-FR-FPN), a Path Aggregation Space Attention FPN (PA-SA-FPN), and a Balance Scale Global Attention FPN (BS-GA-FPN). Their implementation shows a pipeline, meaning gradually enhancing detection performance. We conduct extensive ablation studies to confirm each FPN's effectiveness. Experimental results on five open SAR ship detection datasets (i.e., SSDD [5], Gaofen-SSDD [6], Sentinel-SSDD [6], SAR-Ship-Dataset [7], and HRSID [8]) reveal that Quad-FPN can offer the most superior detection accuracy compared with the other 12 competitive state-of-the-art CNN-based SAR ship detectors. Finally, we also perform the actual ship detection in another two large-scene SAR images from the Sentinel-1 satellite. The satisfactory detection results confirm the excellent migration application capability of Quad-FPN. The software is available online on our website [23].

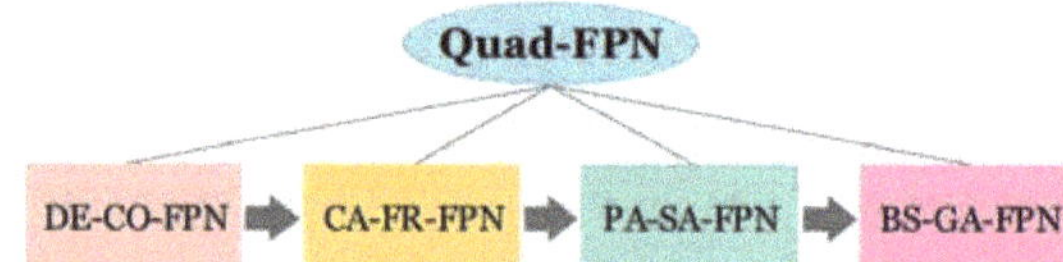

Figure 1. Pipeline structure of Quad-FPN.

The main contributions of this paper are as follows:

1. Quad-FPN is proposed for SAR ship detection.
2. DE-CO-FPN, CA-FR-FPN, PA-SA-FPN, and BS-GA-FPN are designed to improve SAR ship detection performance.
3. Quad-FPN offers the most superior detection accuracy compared with the other 12 competitive state-of-the-art CNN-based SAR ship detectors.

The rest of this paper is arranged as follows. Section 2 introduces Quad-FPN. Section 3 introduces our experiments. Results are shown in Section 4. Ablation studies are presented in Section 5. Finally, a summary of this paper is made in Section 6.

2. Quad-FPN

Quad-FPN is the basis of classical Faster R-CNN [17] and FPN [22], which are both important solutions to handle mainstream detection tasks. Figure 2 shows Quad-FPN's overview. Four basic FPNs, i.e., DE-CO-FPN, CA-FR-FPN, PA-SA-FPN, and BS-GA-FPN, constitute its network architecture. Their implementation presents a pipeline that improves SAR ship detection performance progressively.

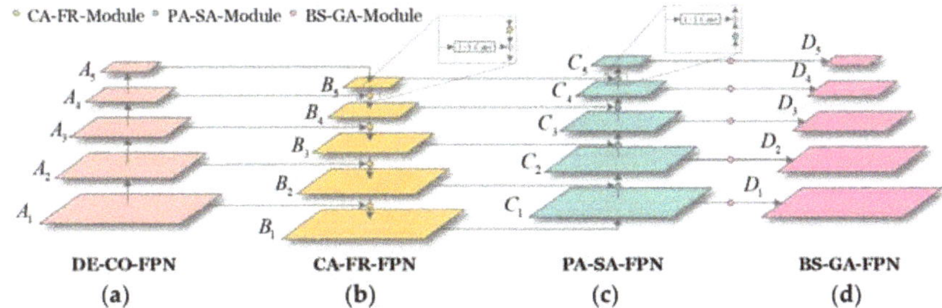

Figure 2. Network architecture of Quad-FPN. (**a**) DE-CO-FPN; (**b**) CA-FR-FPN; (**c**) PA-SA-FPN; and (**d**) BS-GA-FPN.

The overall design idea of Quad-FPN is as follows.

(1) The overall structure of the first two FPNs (DE-CO-FPN and CA-FE-FPN) keeps the same as that of the raw FPN [22], including the sequence of DE-CO-FPN and CA-FE-FPN. In other words, the raw FPN also has two basic sub-FPNs, but they are replaced by our proposed DE-CO-FPN and CA-FE-FPN. Differently, the first sub-FPN in the raw FPN uses the standard convolution, but DE-CO-FPN uses the deformable convolution; the second sub-FPN in the raw FPN uses the simple up-sampling to achieve a feature fusion, but CA-FE-FPN proposes a CA-FR-Module to achieve a feature fusion. DE-CO-FPN's feature maps are from the backbone network, so it is located at the input-end of Quad-FPN. From Figure 2a, DE-CO-FPN realizes the information flow from the bottom to the top. According to the findings in [22], the pyramid top (A_5) has stronger semantic information than its low levels. The semantic information can improve detection performance. Therefore, a top-to-bottom branch in CA-FR-FPN is added to achieve the downward transmission of semantic information. Finally, DE-CO-FPN and CA-FE-FPN form an information interaction loop in which spatial location information and semantic information complement each other.

(2) The design idea of the third FPN (PA-SA-FPN) is inspired from the work of PANET They found that the low-level location information of the pyramid bottom (B_5) was not considered to be transmitted to the top. This might lead to an inaccurate positioning of large objects, so the detection performance of large objects is reduced. Therefore, they added an extra bottom-to-top branch to address this problem. This branch is called PA-FPN in their original reports. Differently, our proposed PA-SA-FPN adds a PA-SA-Module to achieve the feature down-sampling so as to focus on more important spatial features. Finally, CA-FE-FPN and PA-SA-FPN form another information interaction loop in which spatial location information and semantic information complement each other again. Therefore, the overall sequence of DE-CO-FPN, CA-FE-FPN, and PA-SA-FPN is fixed.

(3) The basic outline of Quad-FPN has been determined. BS-GA-FPN is designed to further refine features at each feature level to solve the feature level imbalance of different scale ships. Thus, it is arranged at the output-end of Quad-FPN.

2.1. DEformable COnvolutional FPN (DE-CO-FPN)

The core idea of DE-CO-FPN is that we use the deformable convolution [24] to extract ship features. It contains more useful ship shape information, meanwhile alleviating complex background interferences. Previous work [5–15] mostly adopted the standard or dilated convolutions [25] to extract features. However, the two have limited geometric modeling ability due to their regular kernels. This means that their ability to extract the shape features of multi-scale ships is bound to become poor, causing poor multi-scale detection performance. For inshore ships, the standard and dilated convolutions cannot restrain interferences of port facilities; for ships side-by-side parking at ports, they also cannot eliminate interferences from the nearby ship hull. Thus, to solve this problem, the deformable convolution is used to establish DE-CO-FPN. Figure 3 shows their intuitive comparison. From Figure 3, it is obvious that the deformable convolution can extract ship shape features more effectively; it can suppress the interference of complex backgrounds, especially for more complex inshore scenes. Finally, ships are likely to be separated successfully from complex backgrounds. Thus, this deformable convolution process can be regarded as an extraction of salient objects in various scenes, which plays a role of spatial attention.

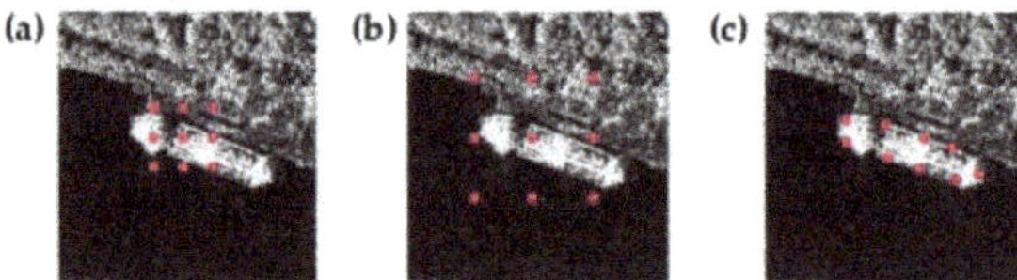

Figure 3. Different convolutions. (a) Standard convolution; (b) dilated convolution; and (c) deformable convolution.

In the deformable convolution, the standard convolution kernel is augmented with offsets $\Delta\mathbf{p}_n$ that are adaptively learned in training to model targets' shape features, i.e.,

$$\mathbf{y}(\mathbf{p}_0) = \sum_{\mathbf{p}_n \in \Re} \mathbf{w}(\mathbf{p}_n) \times \mathbf{x}(\mathbf{p}_0 + \mathbf{p}_n + \Delta\mathbf{p}_n) \tag{1}$$

where $\mathbf{p}_0$ denotes each location, $\Re$ denotes the convolution region, $\mathbf{w}$ denotes the weight parameters, $\mathbf{x}$ denotes the input, $\mathbf{y}$ denotes the output, and $\Delta\mathbf{p}_n$ denotes the learned offsets at the n-th location. It should be noted that compared with standard convolutions, deformable ones' training is in fact time-consuming; it needs more GPU memory. This is because the learned offsets add extra network parameters, increasing networks' complexity. A reasonable fitting of these offsets must be time-consuming. Yet, in this paper, to obtain better accuracy of ships with various shapes, we have not studied this issue deeply for the time being. This problem will be considered with due attention in our future work.

In Equation (1), $\Delta\mathbf{p}_n$ is typically fractional. Thus, we use the bilinear interpolation to ensure the smooth implementation of convolutions, i.e.,

$$\mathbf{x}(\mathbf{p}) = \sum_{\mathbf{q}} G(\mathbf{q}, \mathbf{p}) \times \mathbf{x}(\mathbf{q}) \tag{2}$$

where $\mathbf{p}$ denotes the fraction location to be interpolated, $\mathbf{q}$ denotes all integral spatial locations in the feature map $\mathbf{x}$, and $G(\ )$ denotes the bilinear interpolation kernel defined by

$$G(\mathbf{q}, \mathbf{p}) = g(q_x, p_x) \times g(q_y, p_y), \text{ where } g(a, b) = \max(0, 1 - |a - b|) \tag{3}$$

In experiments, we add another one convolution layer to learn the offsets $\Delta\mathbf{p}_n$. Then, the standard convolution combining $\Delta\mathbf{p}_n$ is performed on the input feature maps. Finally, ship features with rich shape information (A_1, A_2, A_3, A_4, and A_5 in Figure 2a) will be transferred to subsequent FPNs for more operations.

2.2. Content-Aware Feature Reassembly (CA-FR-FPN)

The core idea of CA-FR-FPN is that we design a CA-FR-Module (marked by circle in Figure 2b) to enhance feature transmission benefits when performing the up-sampling multi-level feature fusion. Previous work [5–15] added a feature fusion branch from top to bottom to via feature up-sampling. This feature up-sampling is often completed by the nearest neighbor or bilinear interpolations, but the two means merely consider sub-pixel neighborhoods, which cannot effectively capture the rich semantic information required by dense detection tasks [26], especially for densely distributed small ships. That is, features of small ships are easily diluted because of their poor conspicuousness, leading to feature loss. Thus, to solve this problem, we propose a CA-FR-Module in the up-sampling feature fusion branch from top to bottom to achieve a feature reassembly. It can be aware of important contents in feature maps, and attach importance to key small ship features, thereby improving feature transmission benefits. Figure 2b shows the network architecture of CA-FR-FPN. From Figure 2b, for five-scale levels (B_1, B_2, B_3, B_4, and B_5), four CA-FR-Modules are used for feature reassembly. In practice, CA-FR-Module will complete the task that is similar to the $2\times$ up-sampling operation in essence. Figure 4 shows the implementation process of CA-FR-Module. From Figure 4, there are two basic steps in CA-FR-Module: (1) kernel prediction, and (2) content-aware feature reassembly.

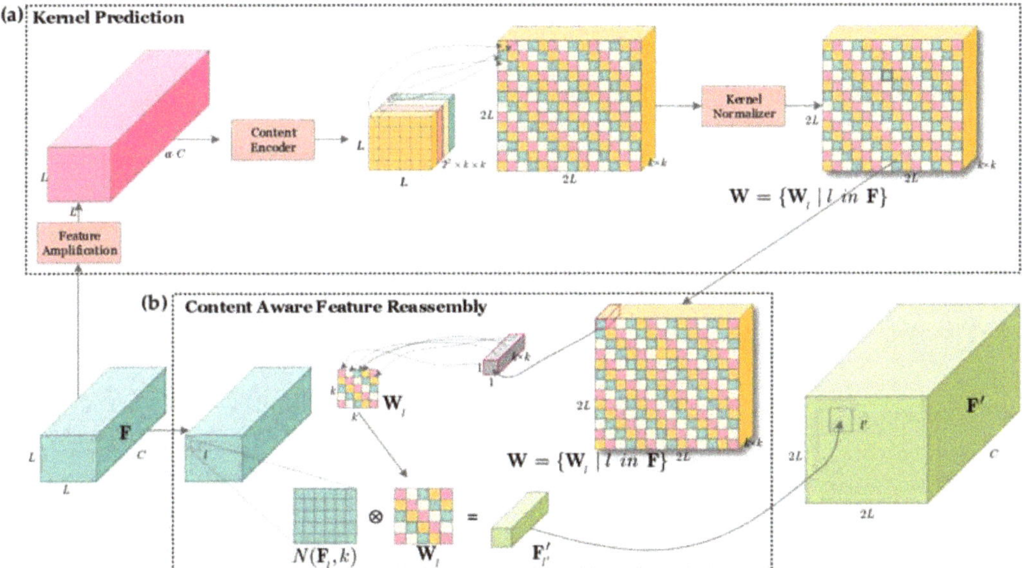

Figure 4. Implementation process of CA-FR-Module in CA-FR-FPN. (**a**) Kernel prediction; (**b**) content-aware feature reassembly.

Step 1: Kernel Prediction

Figure 4a shows the implementation process of the kernel prediction. In Figure 4, the feature maps **F**'s dimension is $L \times L \times C$, where L denotes its size and C denotes its channel width. Overall, the process of the kernel prediction (denoted by ψ) is responsible for generating adaptive feature reassembly kernels $\mathbf{W}_l$ at the original location l, according to the $k \times k$ neighbors of feature maps $\mathbf{F}_l$ through a content-aware manner, i.e.,

$$\mathbf{W}_l = \psi(N(\mathbf{F}_l, k)) \tag{4}$$

where $N(\cdot)$ means the neighbors and $\mathbf{W}_l$ denotes the reassembly kernel.

To enhance the content-aware benefits of the kernel prediction, we first design a convolution layer to amplify the inputted feature maps $\mathbf{F}$ by α times (from C to $\alpha \cdot C$). This convolution layer's kernel number is set to $\alpha \cdot C$, where α is an experimental hyper-parameter that will be studied in Section 5.2.2. Then, we adopt another convolution layer to encode the content of input features so as to obtain reassembly kernels. Here, we set the kernel width as $2^2 \times k \times k$ where 2 is from the requirement of the $2\times$ up-sampling operation. The purpose is to enlarge the size of feature maps to $2L$. Moreover, $k \times k$ is from the $k \times k$ neighbors of feature maps $\mathbf{F}_l$. Afterwards, the content encoded features are reshaped to a $2L \times 2L \times (k \times k)$ dimension via the pixel shuffle means [27]. Finally, each reassembly kernel is normalized by a soft-max function spatially to reflect the weight of each sub-content.

In summary, the above operations can be described by:

$$\mathbf{W}_l = soft-max\left\{shuffle[f_{encode}(f_{amplify}(\mathbf{F}_l))]\right\} \tag{5}$$

where $f_{amplify}$ denotes the feature amplification operation, f_{encode} denotes the content encode operation, $shuffle$ denotes the pixel shuffle means, $soft\text{-}max$ denotes the soft-max function defined by $e^{X_i} / \sum_j e^{X_j}$, and $\mathbf{W}_l$ denotes the generated reassembly kernel.

Step 2: Content-Aware Feature Reassembly

Figure 4b shows the implementation process of the content-aware feature reassembly. Overall, the process of the content-aware feature reassembly (denoted by ϕ) is responsible for generating the final up-sampling feature maps $\mathbf{F}'_{l'}$, i.e.,

$$\mathbf{F}'_{l'} = \phi(N(\mathbf{F}_l, k), \mathbf{W}_l) \tag{6}$$

where k denotes the $k \times k$ neighbors and $\mathbf{W}_l$ denotes the reassembly kernel in Equation (4) that corresponds to the l' location of feature maps after up-sampling from the original l location. For each reassembly kernel $\mathbf{W}_l$, this step will reassemble the features within a local region via the function ϕ in Equation (6). Similar to the standard convolution operation, ϕ can be implemented by a weighted sum. Thus, for a target location l' and the corresponding square region $N(\mathbf{F}_l, k)$ centered at $l = (i, j)$, the reassembly output is described by

$$\mathbf{F}'_{l'} = \sum_{n \in \Re,\, m \in \Re} \mathbf{W}_{l,(n,m)} \times \mathbf{F}_{(i+n,j+m)} \tag{7}$$

where $\Re$ denotes the corresponding square region $N(\mathbf{F}_l, k)$. Moreover, k is set to 5 in our work that is an optimal value followed by [26].

With the reassembly kernel $\mathbf{W}_l$, each pixel in the region $\Re$ of the original location l contributes to the up-sampled pixel l' differently, based on the content of features rather than location distance. Semantic features from the pyramid top will be transferred into the bottom, bringing better transmission benefits. Finally, the pyramid top's features will be fused into the bottom to enhance the feature expression ability of small ships.

2.3. Path Aggregation Space Attention FPN (PA-SA-FPN)

The core idea of PA-SA-FPN is that we add an extra path aggregation branch with a space attention module (PA-SA-Module) (marked by circle in Figure 2c) from the pyramid bottom to the top. Previous work [5–15] often transmitted high-level strong semantic features to the bottom to improve the whole pyramid expressiveness. Yet, the low-level location information from the pyramid bottom was not considered to be transmitted to the top. This can lead to inaccurate positionings of large ship bounding boxes, so the detection performance of large ships is reduced. Thus, we add an extra path aggregation branch (bottom-to-top) to handle this problem. Moreover, to further improve path aggregation benefits, we design a PA-SA-Module to concentrate on important spatial information to avoid interferences of complex port facilities. Figure 2c shows PA-SA-FPN's architecture.

From Figure 2c, the location information of the pyramid bottom is transmitted to the top $(C_1 \rightarrow C_2 \rightarrow C_3 \rightarrow C_4 \rightarrow C_5)$ by the feature down-sampling. In this way, the top semantic features will be enriched with more ship spatial information. This can improve feature expression ability of large ships. Moreover, before the down-sampling, the low-level feature maps are refined by a PA-SA-Module to improve path aggregation benefits [28].

Figure 5 shows the implementation process of PA-SA-Module. In Figure 5, the input feature maps are denoted by $\mathbf{Q}$ and the output ones are denoted by $\mathbf{Q}'$. First, a global average pooling (GAP) [29] is used to obtain the average response in space; a global max pooling (GMP) [29] is used to obtain the maximum response in space. Then, their implementation results are concatenated as the synthetic feature maps, denoted by $\mathbf{S}$. Unlike the previous convolutional block attention module [28], we design a space encoder $f_{space\text{-}encode}$ to encode the space information. It is used to represent the spatial correlation. This can improve spatial attention gains because features in the coding space are more concentrated. Then, the output of $f_{space\text{-}encode}$ is activated by a *sigmod* function to represent each pixel's importance-level in the original space, i.e., an importance-level weight matrix $\mathbf{W}_S$. Finally, an elementwise multiplication is conducted between the original feature maps $\mathbf{Q}$ and the importance-level weight matrix $\mathbf{W}_S$ to obtain the output $\mathbf{Q}'$.

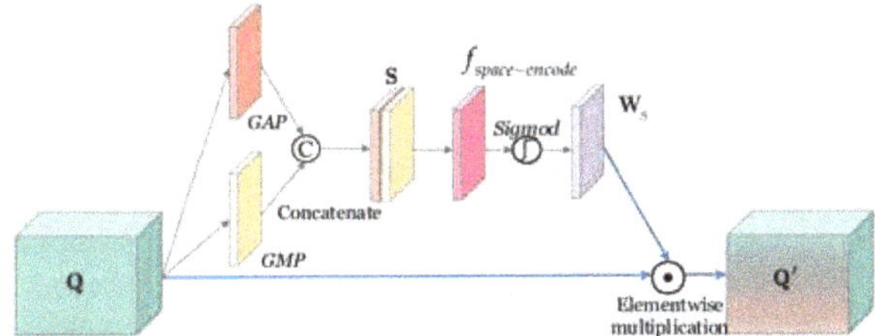

Figure 5. Implementation process of PA-SA-Module in PA-SA-FPN.

In short, the above can be described by

$$\mathbf{Q}' = \mathbf{Q} \odot \mathbf{W}_S \tag{8}$$

where $\mathbf{Q}$ denotes the input feature maps, $\mathbf{Q}'$ denotes the output feature maps, $\odot$ denotes the elementwise multiplication, and $\mathbf{W}_S$ denotes the importance-level weight matrix, i.e.,

$$\mathbf{W}_S = sigmod\left\{ f_{space-encode}(GAP(\mathbf{Q})(c)GMP(\mathbf{Q})) \right\} \tag{9}$$

where GAP denotes the global average-pooling, GMP denotes the global max-pooling, $f_{space\text{-}encode}$ denotes the space encoder, © denotes the concatenation operation, and *sigmod* is an activation function defined by $1/(1 + e^{-x})$.

Finally, the feature pyramid will be stronger when possessing both the top-to-bottom branch and bottom-to-top branch. Each level has rich spatial location information and abundant semantic information, which help improve large ships' detection performance.

2.4. Balance Scale Global Attention FPN (BS-GA-FPN)

The core idea of BS-GA-FPN is that we further refine features from each feature level in the pyramid, to address the feature level imbalance of different scale ships. SAR ships often present different characteristics at different levels in the pyramid, i.e., the existence of multi-scale ship feature differences. Due to the difference of resolutions, the difference of satellite shooting distances, and different slicing methods, there are many scales of ships in the existing SAR ship datasets. E.g., for SSDD, the smallest ship pixel size is 7×7 while the biggest one is 211×298. Such huge size gap results in large ship feature differences, which makes it very difficult to detect them. In the computer vision community, Pang et al. [30] found that such feature level imbalance may weaken the feature expression capacity of FPN, but previous work [5–15] in the SAR ship detection community was not aware of this problem. Thus, to handle this problem, we design a BS-GA-Module to further process pyramid features to recover a balanced BS-GA-FPN. Implementation process of BS-GA-Module consists of four steps: (1) feature pyramid resizing, (2) balanced multi-scale feature fusion, (3) global attention (GA) refinement, and (4) feature pyramid recovery, as in Figure 6.

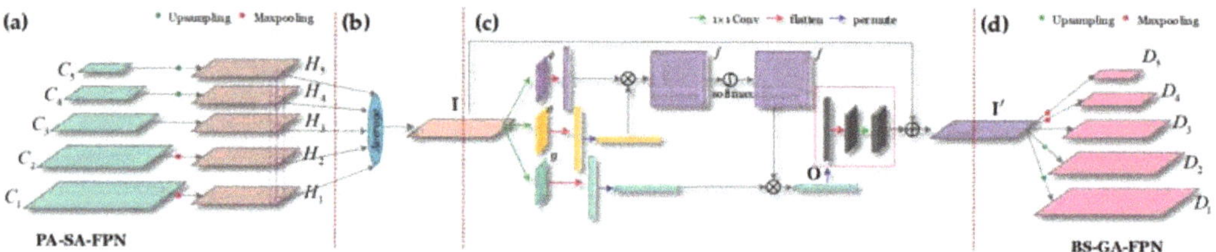

Figure 6. Implementation process of BS-GA-Module. (**a**) Feature pyramid resizing; (**b**) balanced multi-scale feature fusion; (**c**) GA refinement; and (**d**) feature pyramid recovery.

Step 1: Feature Pyramid Resizing

Figure 6a shows the graphical description of the feature pyramid resizing. In Figure 6a, in the PA-SA-FPN, features maps at different levels are denoted by C_1, C_2, C_3, C_4, and C_5. To facilitate the fusion of balanced features to preserve their semantic hierarchy at the same time, we resize each detection scale (C_1, C_2, C_3, C_4, and C_5) to a unified resolution, by a max-pooling or up-sampling. Here, C_3 is selected as this unified resolution level because it locates in the middle of the pyramid. It can maintain a trade-off between top semantic information and bottom spatial information. Finally, the above can be described by

$$H_1 = MaxPool^{4\times}(C_1), \; H_2 = MaxPool^{2\times}(C_2), \; H_3 = C_3, \; H_4 = UpSampling^{2\times}(C_4), \; H_5 = UpSampling^{4\times}(C_5) \quad (10)$$

where H_1, H_2, H_3, H_4, and H_5 are the resized feature maps from the original ones, $UpSampling^{n\times}$ denotes the n times up-sampling, and $MaxPool^{n\times}$ denotes the n times max-pooling.

Step 2: Balanced Multi-Scale Feature Fusion

Figure 6b shows the graphical description of the balanced multi-scale feature fusion. After obtaining feature maps with the same unified resolution, the balanced multi-scale feature fusion is executed by

$$\mathbf{I}(i,j) = \frac{1}{5}\sum_{k=1}^{5} H_k(i,j) \quad (11)$$

where k denotes the k-th detection level, (i, j) denotes the spatial location of feature maps, and $\mathbf{I}$ denotes the output integrated features. From Equation (11), the features from each scale (H_1, H_2, H_3, H_4, and H_5) are uniformly fused as the output $\mathbf{I}$ (a mean operation). Here, the average operation fully reflects the balanced idea of SAR ship scale feature fusion.

Finally, the output $\mathbf{I}$ with condensed multi-scale information will contain balanced semantic features of various resolutions. In this way, big ship features and small ones can complement each other to facilitate the information flow.

Step 3: GA Refinement

To make features from different scales become more discriminative, we also propose a GA refinement mechanism to further refine balanced features in Equation (11). This can enhance their global response ability. That is, the network will pay more attention to important spatial global information (feature self-attention), as in Figure 6c.

The GA refinement can be described by

$$O_i = \frac{1}{\xi(\mathbf{I})} \times \sum_{\forall j} f(I_i, I_j) \times g(I_j) \tag{12}$$

where I_i denotes the input at the i-th location, O_i denotes the output at the i-th location, $f(\cdot)$ is a function used to calculate the similarity between the location I_i and I_j, $g(\cdot)$ is a function to characterize the feature representation at the j-th location, and $\xi(\cdot)$ denotes a normalized coefficient (the input overall response). The i-th location information denotes the current location's response, and the j-th location information denotes the global response.

In Equation (12), $g(\cdot)$ can be regarded as a linear embedding,

$$g(I_j) = W_g I_j \tag{13}$$

where Wg is a weight matrix to be learned, and we use a 1×1 convolutional layer to obtain this weight matrix during training.

Furthermore, one simple extension of the Gaussian function is to compute similarity $f(\cdot)$ in an embedding space,

$$f(I_i, I_j) = e^{\theta(I_i)^T \phi(I_j)} \tag{14}$$

where $\theta(I_i) = W_\theta I_i$ and $\phi(I_j) = W_\phi I_j$ are two embeddings. W_θ and W_ϕ are the weight matrixes to be learned that are both achieved by other two 1×1 convolutional layers.

As above, the normalized coefficient $\xi(\cdot)$ is set to

$$\xi(\mathbf{I}) = \sum_{\forall j} f(I_i, I_j) \tag{15}$$

Finally, the whole GA refinement is instantiated as:

$$O_i = \left(e^{\theta(I_i)^T \phi(I_j)} \times W_g I_j \right) / \sum_{\forall j} e^{\theta(I_i)^T \phi(I_j)} \tag{16}$$

where $e^{\theta(I_i)^T \phi(I_j)} / \sum_{\forall j} e^{\theta(I_i)^T \phi(I_j)}$ can be achieved by a soft-max function.

Figure 6c shows the graphical description of the above GA refinement. From Figure 6c, two 1×1 convolutional layers are used to compute ϕ and θ. Then, by the matrix multiplication $\theta^T \phi$, the similarity f is obtained. One 1×1 convolutional layer is used to characterize the representation of the features g. Finally, f with a soft-max function multiplies by g to obtain the feature self-attention output $\mathbf{O} = \{O_i \mid i \text{ in } \mathbf{I}\}$. Finally, the feature self-attention output $\mathbf{O}$ is further processed by one 1×1 convolutional layer (marked in a dotted box). The purpose is to make $\mathbf{O}$ match the dimension of the original input $\mathbf{I}$ to facilitate follow-up element-wise adding. This is similar to the residual/skip connections of ResNet. Consequently, the refined features $\mathbf{I}'$ combining the feature self-attention information are achieved, which will be further processed in the subsequent steps, i.e.,

$$\mathbf{I}' = W_O \mathbf{O} + \mathbf{I} \tag{17}$$

where W_O is also a weight matrix to be learned, and another 1×1 convolutional layer can be used to obtain it during training.

In essence, the GA refinement can directly capture long-range dependence of each location (global response) by calculating the interaction between two different arbitrary positions. It is equivalent to constructing a convolutional kernel with the same size as the feature map $\mathbf{I}$, to maintain more useful ship information, making feature maps more discriminative. More detailed theories about this global attention can be found in [31].

Step 4: Feature Pyramid Recovery

Figure 6d shows the graphical description of the feature pyramid recovery. From Figure 6d, the refined features $\mathbf{I}'$ are resized again through using the similar but reverse procedure of Equation (10) to recover a balanced feature pyramid, i.e.,

$$D_1 = UpSampling^{4\times}(\mathbf{I}'),\ D_2 = UpSampling^{2\times}(\mathbf{I}'),\ D_3 = \mathbf{I}',\ D_4 = MaxPool^{2\times}(\mathbf{I}'),\ D_5 = MaxPool^{4\times}(\mathbf{I}') \tag{18}$$

where D_1, D_2, D_3, D_4, and D_5 denote the recovered feature maps at different levels after ship scale balance operations. They reconstruct the final network architecture of BS-GA-FPN. Ultimately, D_1, D_2, D_3, D_4, and D_5 in BS-GA-FPN will possess more multi-scale balanced features that will be used to be responsible for the final ship detection.

3. Experiments

Our experiments are run on a personal computer with i9-9900K CPU and RTX2080Ti GPU based on Pytorch. Quad-FPN and the other 12 competitive SAR ship detectors are implemented under the MMDetection toolbox [32] to ensure the comparison fairness.

3.1. Experimental Datasets

(1) **SSDD:** SSDD is the first open SAR ship detection dataset, proposed by Li et al. [5] in 2017. There are 1160 SAR images with 500×500 average image size in SSDD from Sentinel-1, TerraSAR-X, and RadarSat-2. SAR ships in SSDD are provided with various resolutions from 1m to 10m, and HH, HV, VV, and VH polarizations. We set the ratio of the training set and the test set to 8:2. Here, image names with the index suffix of 1 and 9 are selected as the test set, and the others as the training set.

(2) **Gaofen-SSDD:** Gaofen-SSDD was constituted in [6] to make up for the shortcoming of insufficient samples in SSDD. There are 20,000 images with 160×160 image size in Gaofen-SSDD from Gaofen-3. SAR ships in Gaofen-SSDD are provided with various resolutions from 5 m to 10 m, and HH, HV, VV, and VH polarizations. Same as [6], the ratio of the training set, validation set, and the test set is 7:2:1 by a random selection.

(3) **Sentinel-SSDD:** Sentinel-SSDD was constituted in [6] to make up for the shortcoming of insufficient sample number in SSDD. There are 20,000 images with 160×160 image size in Sentinel-SSDD from Sentinel-1. SAR ships in Sentinel-SSDD are provided with resolutions from 5 m to 20 m, and HH, HV, VV, and VH polarizations. Same as [6], the ratio of the training set, validation set, and the test set is 7:2:1 by a random selection.

(4) **SAR-Ship-Dataset:** SAR-Ship-Dataset was released by Wang et al. [7] in 2019. There are 43,819 images with 256×256 image size in SAR-Ship-Dataset from Sentinel-1 and Gaofen-3. SAR ships in Sentinel-SSDD are provided with resolutions from 5 m to 20 m, and HH, HV, VV, and VH polarizations. Same as their original reports in [7], the ratio of the training set, validation set, and the test set is 7:2:1 by a random selection.

(5) **HRSID:** HRSID was released by Wei et al. [8] in 2020. There are 5604 images with 800×800 image size in HRSID from Sentinel-1 and TerraSAR-X. SAR ships in HRSID are provided with resolutions from 0.1 m to 3 m, and HH, HV, and VV polarizations. Same as its original reports in [8], the ratio of the training set and the test set is 13:7 according to its default configuration files.

3.2. Experimental Details

ResNet-50 with pretraining on ImageNet [33] serves as Quad-FPNs' backbone network. Images in SSDD, Gaofen-SSDD, Sentinel-SSDD, SAR-Ship-Dataset, and HRSID are resized as the 512×512, 160×160, 160×160, 256×256, and 800×800 image size for training. We train Quad-FPN for 12 epochs with a batch size of 2, due to the limited GPU memory. Stochastic gradient descent (SGD) [34] serves as the optimizer with a 0.1 learning rate, a 0.9 momentum, and a 0.0001 weight decay. Moreover, the learning rate is reduced by 10 times per epoch from 8-epoch to 11-epoch to ensure an adequate loss reduction. Followed by Wei et al. [12], a soft non-maximum suppression (Soft-NMS) [35] algorithm is used to suppress duplicate detections with an intersection over union (IOU) threshold of 0.5.

3.3. Loss Function

Followed by Cui et al. [13], the cross entropy (CE) serves as the classification loss L_{cls},

$$L_{cls} = -\frac{1}{N}\sum_{i=1}^{N} p_i \log(p_i^*) + (1 - p_i)\log(1 - p_i^*) \tag{19}$$

where p_i denotes the predictive class probability, p_i^* denotes the ground truth class label, and N denotes the prediction number. The smooth_{L1} serves as the regression loss L_{reg},

$$L_{reg} = \frac{1}{N}\sum_{i=1}^{N} p_i^* \text{smooth}_{L1}(t_i - t_i^*), where\ \text{smooth}_{L1}(x) = \begin{cases} 0.5x^2\ if\ |x| < 1 \\ |x| - 0.5\ otherwise \end{cases} \tag{20}$$

where t_i denotes the predictive bounding box and t_i^* denotes the ground truth box.

3.4. Evaluation Indices

Evaluation indices from the PASCAL dataset [5] are adopted by this paper, including the recall (r), precision (p), and mean average precision (mAP) [36], i.e.,

$$r = TP/(TP + FN),\ p = TP/(TP + FP),\ \text{mAP} = \int_0^1 p(r) \times dr \tag{21}$$

where TP denotes the number of true positives, FN denotes that of false negatives, FP denotes that of false positives, and $p(r)$ denotes the precision-recall curve. In this paper, mAP measures the final detection accuracy because it considers both precision and recall.

Moreover, the frames per second (FPS) is used to measure the detection speed, which is defined by $1/t$, where t refers to the time to detect an image, whose unit is the second (s).

4. Results

4.1. Quantitative Results on Five Datasets

Tables 1–5 show the quantitative comparison with the other 12 competitive state-of-the-art CNN-based SAR ship detectors, on SSDD, Gaofen-SSDD, Sentinel-SSDD, SAR-Ship-Dataset, and HRSID. From Tables 1–5, one can clearly find that:

1. On SSDD, Quad-FPN offers the best accuracy (95.29% mAP on the entire scenes). The second-best one is 92.27% mAP in the entire scenes from DCN [24], but it is still lower than Quad-FPN by ~3% mAP, showing the best detection performance of Quad-FPN.
2. On Gaofen-SSDD, Quad-FPN offers the best accuracy (92.84% mAP on the entire scenes). The second-best one is 91.35% mAP in the entire scenes from Free-Anchor, but it is still lower than Quad-FPN by ~1.5% mAP, showing the best detection performance of Quad-FPN.
3. On Sentinel-SSDD, Quad-FPN offers the best accuracy (95.20% mAP on the entire scenes). The second-best one is 94.31% mAP in the entire scenes from Free-Anchor, but it is still lower than Quad-FPN by ~1% mAP, showing the best detection performance of Quad-FPN.
4. On SAR-Ship-Dataset, Quad-FPN offers the best accuracy (94.39% mAP on the entire scenes). The second-best one is 93.70% mAP in the entire scenes from Free-Anchor, but it is still lower than Quad-FPN by ~1% mAP, showing the best detection performance of Quad-FPN.
5. On HRSID, Quad-FPN offers the best detection accuracy (86.12% mAP on the entire scenes). The second-best one is 83.72% mAP in the entire scenes from Guided Anchoring, but it is still lower than Quad-FPN by ~3.5% mAP.

6. Furthermore, for Quad-FPN and the other 12 methods, the detection accuracies of inshore scenes are all lower than that of offshore scenes. This is in line with common sense because the former has more complex backgrounds than the latter.

7. For the more complex inshore scenes, the detection accuracy advantage of Quad-FPN is more obvious than the other 12 methods. Specifically, Quad-FPN offers an accuracy of 84.68% mAP on the SSDD's inshore scenes, superior to the second-best DCN [24] by ~10% mAP; it offers an accuracy of 85.68% mAP on the Gaofen-SSDD's inshore scenes, superior to the second-best Free-Anchor by ~4% mAP; it offers an accuracy of 84.68% mAP on the Sentinel-SSDD's inshore scenes, superior to the second-best Free-Anchor by ~5% mAP; it offers an accuracy of 83.93% mAP on the SAR-Ship-Dataset's inshore scenes, superior to the second-best Double-Head R-CNN by ~2% mAP; and it offers an accuracy of 70.80% mAP on the HRSID's inshore scenes, superior to the second-best Guided Anchoring by ~7% mAP. Thus, Quad-FPN seems to be robust for background interferences because the deformable convolution can suppress the interference of complex backgrounds, especially for inshore scenes.

8. The r values of the other 12 methods are lower than Quad-FPN, perhaps from their poor small ship detection performance. The p values of Quad-FPN are sometimes lower than others. Thus, an appropriate score threshold can be further considered in the future to make a trade-off between missed detections and false alarms.

9. To be honest, Quad-FPN sacrifices speed due to the network's high-complexity. Yet, it is also important to further improve the accuracy, e.g., the precision strike of military targets. In the future, we will make a trade-off between accuracy and speed.

4.2. Qualitative Results on Five Datasets

Figures 7–11 show the qualitative results on SSDD, Gaofen-SSDD, Sentinel-SSDD, SAR-Ship-Dataset, and HRSID. Here, we only compare Quad-FPN with the second-best detector, due to limited pages.

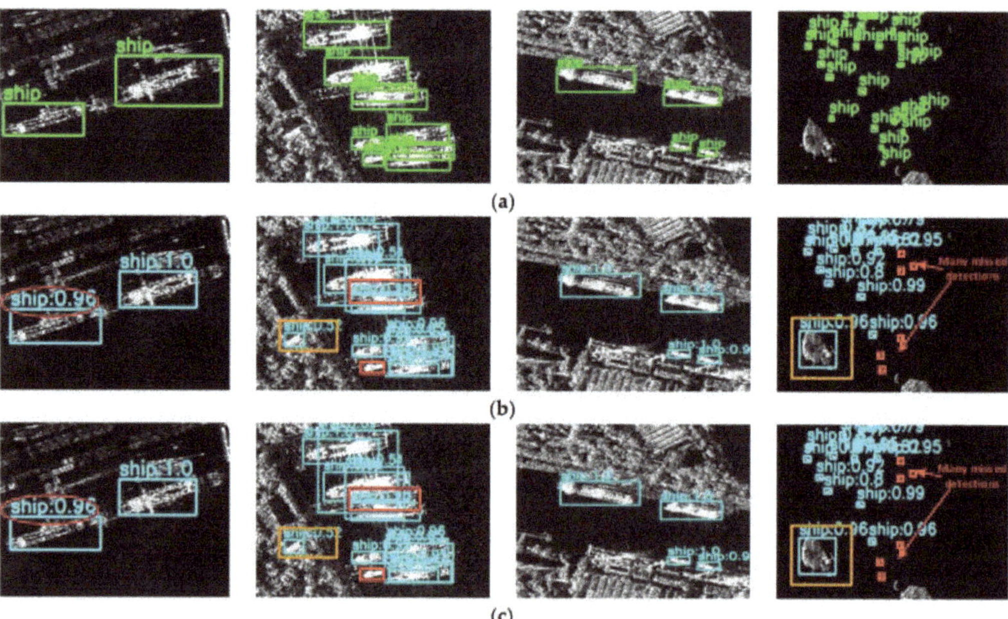

Figure 7. SAR ship detection results on SSDD. (**a**) Ground truths; (**b**) detection results of the second-best DCN [24]; and (**c**) detection results of the first-best Quad-FPN. Missed detections are marked by red boxes; false alarms are marked by orange boxes.

Table 1. Quantitative evaluation indices comparison with the other 12 state-of-the-art CNN-based detectors on SSDD.

No.	Method	Entire Scenes			Inshore Scenes			Offshore Scenes			FPS
		r (%)	p (%)	mAP (%)	r (%)	p (%)	mAP (%)	r (%)	p (%)	mAP (%)	
1	Faster R-CNN [22]	90.44	87.08	89.74	75.00	68.98	71.39	97.58	96.03	97.37	11.87
2	PANET [37]	91.91	86.81	91.15	77.33	67.17	72.92	98.66	97.09	98.48	11.65
3	Cascade R-CNN [21]	90.81	94.10	90.50	74.42	84.21	72.75	98.39	98.12	98.32	9.46
4	Double-Head R-CNN [38]	91.91	86.96	91.10	78.49	68.18	74.97	98.12	96.82	97.82	6.52
5	Grid R-CNN [39]	89.71	87.77	88.92	70.93	68.16	67.40	98.39	97.08	98.13	9.18
6	DCN [24]	93.01	86.20	92.27	79.65	64.93	74.86	99.19	98.14	99.12	11.28
7	Guided Anchoring [40]	90.44	94.62	90.01	73.26	86.90	71.59	98.39	97.60	98.22	9.64
8	Free-Anchor [41]	92.65	72.31	91.04	80.81	45.42	72.37	98.12	93.35	97.72	12.76
9	HR-SDNet [12]	90.99	96.49	90.82	74.42	90.78	73.65	98.66	98.66	98.59	5.79
10	DAPN [13]	91.36	85.54	90.56	77.91	64.11	73.22	97.58	97.58	97.41	12.22
11	SER Faster R-CNN [14]	92.28	86.11	91.52	79.07	66.34	74.56	98.39	96.83	98.26	11.64
12	ARPN [15]	90.62	85.44	89.85	75.00	63.86	70.70	97.85	97.07	97.70	12.15
13	**Quad-FPN (Ours)**	95.77	89.52	**95.29**	87.79	74.75	**84.68**	99.46	97.37	**99.38**	11.37

The best detector is bold and the second-best is underlined.

Table 2. Quantitative evaluation indices comparison with the other 12 state-of-the-art CNN-based detectors on Gaofen-SSDD.

No.	Method	Entire Scenes			Inshore Scenes			Offshore Scenes			FPS
		r (%)	p (%)	mAP (%)	r (%)	p (%)	mAP (%)	r (%)	p (%)	mAP (%)	
1	Faster R-CNN [22]	86.71	88.13	84.25	79.97	77.63	73.99	89.15	92.19	87.56	21.44
2	PANET [37]	85.46	87.40	83.12	76.83	76.25	71.38	88.60	91.62	87.03	21.94
3	Cascade R-CNN [21]	89.06	90.92	87.12	80.73	82.50	75.84	92.08	93.97	90.89	17.40
4	Double-Head R-CNN [38]	87.28	88.47	85.03	80.23	78.06	74.83	89.84	92.46	88.38	9.04
5	Grid R-CNN [39]	87.11	86.73	84.54	79.85	73.72	73.01	89.75	91.98	88.20	15.27
6	DCN [24]	87.78	86.91	85.16	80.35	75.50	74.29	90.48	91.36	88.72	19.42
7	Guided Anchoring [40]	91.47	91.72	89.91	85.14	85.14	80.87	93.78	94.12	92.85	17.98
8	Free-Anchor [41]	93.39	74.74	91.35	87.78	54.67	81.14	95.42	85.21	94.38	24.49
9	HR-SDNet [12]	91.04	91.59	89.43	82.62	82.83	78.18	94.10	94.79	93.16	7.93
10	DAPN [13]	88.55	87.01	86.09	80.35	75.77	74.62	91.53	91.32	89.88	21.25
11	SER Faster R-CNN [14]	85.70	87.40	83.27	77.71	76.74	71.93	88.60	91.45	86.97	20.81
12	ARPN [15]	91.54	90.62	89.73	78.51	76.90	76.73	90.64	92.38	92.23	21.05
13	**Quad-FPN (Ours)**	95.37	75.58	**92.84**	94.21	59.55	**85.68**	95.79	83.62	**94.54**	21.81

The best detector is bold and the second-best is underlined.

Table 3. Quantitative evaluation indices comparison with the other 12 state-of-the-art CNN-based detectors on Sentinel-SSDD.

No.	Method	Entire Scenes			Inshore Scenes			Offshore Scenes			FPS
		r (%)	p (%)	mAP (%)	r (%)	p (%)	mAP (%)	r (%)	p (%)	mAP (%)	
1	Faster R-CNN [22]	91.63	88.31	90.64	76.70	67.82	70.14	96.89	96.42	96.51	23.07
2	PANET [37]	92.34	88.16	91.38	78.57	67.64	71.47	97.19	96.49	96.91	22.59
3	Cascade R-CNN [21]	91.67	91.76	90.94	76.19	76.06	71.17	97.13	97.30	96.91	15.31
4	Double-Head R-CNN [38]	92.16	90.09	91.35	78.06	71.38	72.24	97.13	97.30	96.85	7.76
5	Grid R-CNN [39]	91.28	86.16	90.16	76.53	62.33	67.77	96.47	96.47	96.16	7.59
6	DCN [24]	92.83	87.44	91.79	80.44	65.42	72.55	97.19	96.95	96.86	9.57
7	Guided Anchoring [40]	92.87	93.83	92.20	80.95	83.80	76.81	97.07	97.24	96.86	9.35
8	Free-Anchor [41]	95.00	83.50	94.31	86.22	58.14	79.77	98.08	96.52	**97.98**	24.82
9	HR-SDNet [12]	93.71	92.04	92.97	82.48	76.50	76.45	97.66	97.96	97.55	8.32
10	DAPN [13]	91.54	88.60	90.55	76.36	68.03	69.97	96.89	96.71	96.51	21.24
11	SER Faster R-CNN [14]	92.07	87.76	91.08	77.89	67.16	71.32	97.07	96.09	96.70	22.80
12	ARPN [15]	92.60	89.06	91.48	84.63	78.72	77.62	97.69	96.80	97.19	21.10
13	**Quad-FPN (Ours)**	96.28	84.13	**95.20**	92.52	61.40	**84.68**	97.60	96.00	97.30	22.03

The best detector is bold and the second-best is underlined.

Table 4. Quantitative evaluation indices comparison with the other 12 state-of-the-art CNN-based detectors on SAR-Ship-Dataset.

No.	Method	Entire Scenes			Inshore Scenes			Offshore Scenes			FPS
		r (%)	p (%)	mAP (%)	r (%)	p (%)	mAP (%)	r (%)	p (%)	mAP (%)	
1	Faster R-CNN [22]	93.24	86.85	91.73	86.80	69.55	79.47	95.36	93.85	94.65	23.74
2	PANET [37]	93.44	86.89	92.01	87.15	70.37	80.57	95.52	93.48	94.81	23.24
3	Cascade R-CNN [21]	93.48	90.50	92.27	86.12	77.49	80.93	95.90	95.23	95.31	16.80
4	Double-Head R-CNN [38]	94.16	88.49	92.91	88.18	72.40	82.12	96.13	94.86	95.56	9.06
5	Grid R-CNN [39]	93.08	85.18	91.48	85.50	65.54	77.62	95.58	93.43	94.86	16.39
6	DCN [24]	93.25	86.25	91.80	87.08	68.15	80.21	95.29	93.74	94.60	20.62
7	Guided Anchoring [40]	93.80	92.59	92.73	86.12	82.54	81.74	96.33	96.03	95.79	17.41
8	Free-Anchor [41]	94.91	83.99	93.70	88.04	64.05	81.61	97.17	92.60	**96.69**	24.64
9	HR-SDNet [12]	93.29	92.11	92.29	86.19	80.80	81.88	95.63	96.11	95.14	7.88
10	DAPN [13]	93.34	87.28	91.97	86.94	70.08	80.34	95.45	94.21	94.81	21.53
11	SER Faster R-CNN [14]	93.58	86.78	92.18	87.01	69.37	80.24	95.74	93.83	95.11	22.84
12	ARPN [15]	92.01	88.11	91.35	87.92	72.77	81.14	95.00	94.79	95.10	21.52
13	**Quad-FPN (Ours)**	96.10	77.55	**94.39**	92.37	55.01	**83.93**	97.33	88.95	96.59	22.96

The best detector is bold and the second-best is underlined.

Table 5. Quantitative evaluation indices comparison with the other 12 state-of-the-art CNN-based detectors on HRSID.

No.	Method	Entire Scenes			Inshore Scenes			Offshore Scenes			FPS
		r (%)	p (%)	mAP (%)	r (%)	p (%)	mAP (%)	r (%)	p (%)	mAP (%)	
1	Faster R-CNN [22]	81.97	81.45	80.66	65.51	65.55	60.10	97.17	95.93	97.09	14.05
2	PANET [37]	82.98	81.28	81.59	67.55	65.83	61.75	97.24	95.68	97.14	13.00
3	Cascade R-CNN [21]	82.24	87.14	81.27	66.03	74.50	61.93	97.21	97.52	97.14	11.14
4	Double-Head R-CNN [38]	83.36	81.73	82.09	68.36	66.37	63.14	97.21	96.17	97.13	7.05
5	Grid R-CNN [39]	80.91	82.82	79.42	63.60	66.85	57.25	96.88	96.85	96.78	11.07
6	DCN [24]	83.47	81.46	82.09	68.32	65.82	62.39	97.47	96.28	97.39	12.66
7	Guided Anchoring [40]	84.62	90.41	83.72	70.64	81.22	66.99	97.53	97.82	97.46	10.49
8	Free-Anchor [41]	84.39	65.75	81.84	70.05	45.01	60.01	97.63	94.61	97.53	15.76
9	HR-SDNet [12]	82.34	88.89	81.52	65.96	77.69	62.45	97.47	97.69	97.41	6.74
10	DAPN [13]	83.37	80.50	81.84	68.50	64.78	62.29	97.11	95.62	97.01	12.93
11	SER Faster R-CNN [14]	82.97	80.05	81.51	67.41	63.62	61.41	97.34	95.87	97.23	13.59
12	ARPN [15]	83.83	85.74	81.76	68.11	70.74	63.52	97.04	97.27	97.35	12.80
13	**Quad-FPN (Ours)**	87.29	87.96	**86.12**	75.78	77.31	**70.80**	97.92	97.57	**97.86**	13.35

The best detector is bold and the second-best is underlined.

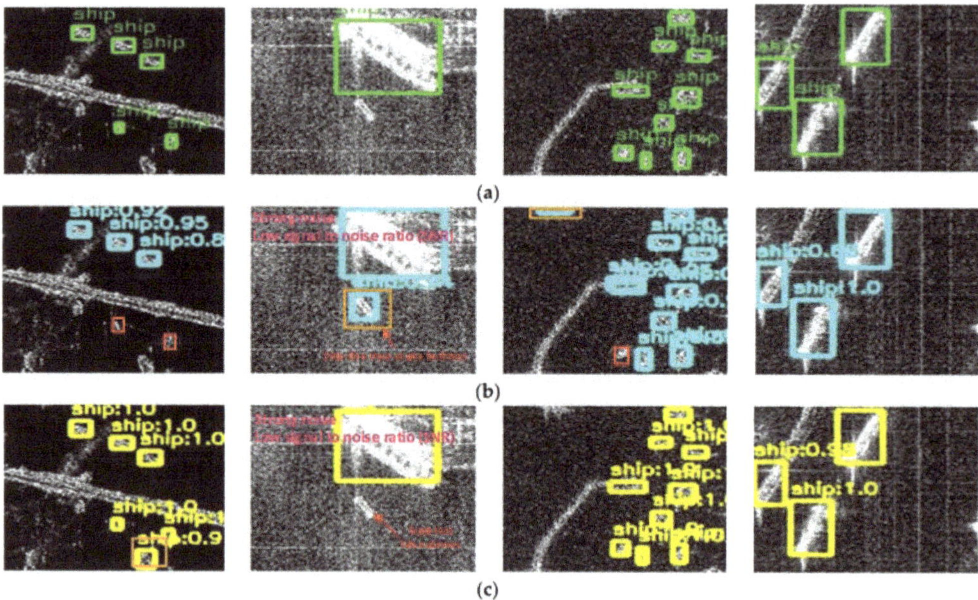

Figure 8. SAR ship detection results on Gaofen-SSDD. (a) Ground truth; (b) detection results of the second-best Free-Anchor [41]; and (c) detection results of the first-best Quad-FPN. Missed detections are marked by red boxes; false alarms are marked by orange boxes.

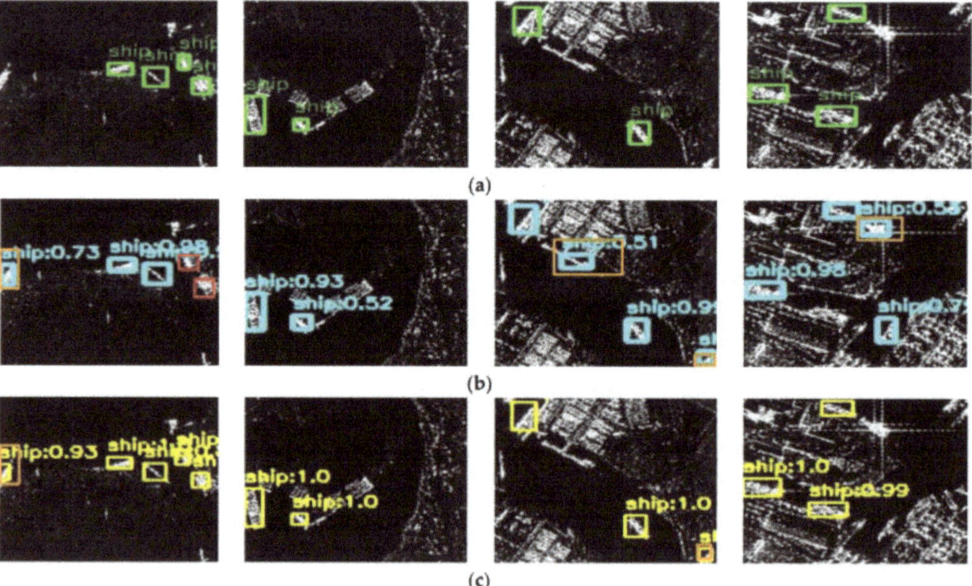

Figure 9. SAR ship detection results on Sentinel-SSDD. (a) Ground truths; (b) detection results of the second-best Free-Anchor [41]; and (c) detection results of the first-best Quad-FPN. Missed detections are marked by red boxes; false alarms are marked by orange boxes.

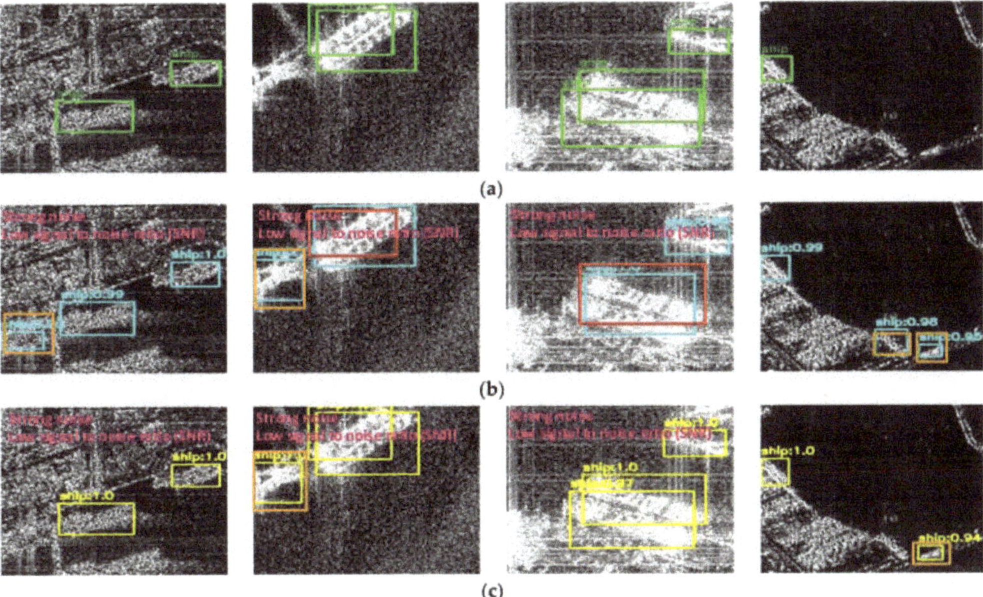

Figure 10. SAR ship detection results on SAR-Ship-Dataset. (**a**) Ground truths; (**b**) detection results of the second-best Free-Anchor [41]; and (**c**) detection results of the first-best Quad-FPN. Missed detections are marked by red boxes; false alarms are marked in orange.

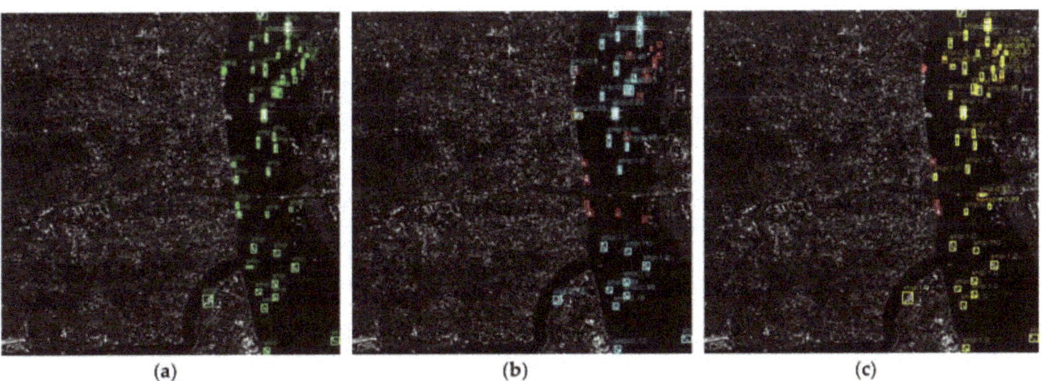

Figure 11. SAR ship detection results on HRSID. (**a**) Ground truths; (**b**) detection results of the second-best Guided Anchoring [40]; and (**c**) detection results of the first-best Quad-FPN. Missed detections are marked by red boxes; false alarms are marked by orange boxes.

Taking SSDD in Figure 7 as an example, we can draw the following conclusions:

1. Quad-FPN can successfully detect various SAR ships with different sizes under various backgrounds. This shows its excellent detection performance with excellent scale-adaptation and scene-adaptation. Compared with the second-best CNN-based ship detector DCN [24], Quad-FPN can improve the detection confidence scores. For example, in the first detection sample of Figure 7, Quad-FPN increases the confidence score from 0.96 to 1.0. This can show Quad-FPN's higher credibility.

2. Quad-FPN can suppress some false alarms from complex inshore facilities. For example, in the second detection sample of Figure 7, one land false alarm is removed by Quad-FPN. This shows Quad-FPN's better scene-adaptability.

3. Quad-FPN can avoid some missed detections of densely arranged ships and small ships. For example, in the second sample of Figure 7, small ships densely parked at ports are detected again by Quad-FPN. This is because the adopted deformable convolution in DE-CO-FPN can alleviate the negative influence from the hull of a nearby ship. In the fourth sample of Figure 7, many small ships are detected successfully again by Quad-FPN, but DCN failed most of them. This is because CA-FR-FPN can transmit more abundant semantic information from the pyramid top to the bottom, to improve the expression capacity of small ship features. This shows Quad-FPN's better detection capacity of both inshore ships and small ones.

4. Moreover, from the third sample of Figure 7, ships with different scales on the same SAR image are detected at the same time. This is because the proposed BS-GA-GPN can balance the feature differences of different sizes of ships, showing Quad-FPN's excellent scale-adaptability.

Moreover, from the detection results of the second sample on Gaofen-SSDD in Figure 8, Quad-FPN can remove false alarms from ship-like man-made facilities, meanwhile successfully detecting the ship moored at port, even under the strong speckle noise interference, or rather low signal to noise ratio (SNR). This shows Quad-FPN has both keen judgment merits and robust anti-noise performance. Similarly, the detection results of the first three samples on SAR-Ship-Dataset in Figure 10 can also reveal its excellent anti-noise performance. Finally, from the detection results of the third sample on SAR-Ship-Dataset in Figure 10, a large ship parking at port is detected by Quad-FPN again. This is because PA-SA-FPN can transmit the low-level location information from the pyramid bottom to the pyramid top, which can bring more accurate positionings of large ship bounding boxes. Correspondingly, the feature learning benefits of large ships are enhanced, thereby avoiding their missed detections. Given the above, Quad-FPN offers state-of-the-art SAR ship detection performance.

4.3. Large-Scene Application in Sentinel-1 SAR Images

We conduct the actual ship detection in another two large-scene Sentinel-1 SAR images to confirm the good migration capability of Quad-FPN. Figure 12 shows the coverage areas of the two large-scene Sentinel-1 SAR images. The two areas are both the world's major shipping routes, so they are selected. Table 6 shows their descriptions. From Table 6, the VV polarization SAR images are selected given that ships generally exhibit higher backscattering values in VV polarization [42]. In addition, the interferometric wide-swath (IW) mode of Sentinel-1 is selected specifically because it is the main mode to acquire data in areas of maritime surveillance interest [42]. The ship ground truths are annotated by SAR experts using the automatic identification system (AIS) and Google Earth. This can provide a more reliable performance evaluation. These two SAR images are resized as 24,000 × 16,000 image size, respectively. Then, followed by [43], they are cut into 800 × 800 small sub-images directly for training and testing because of the limited GPU memory. Finally, they are inputted into Quad-FPN for the actual SAR ship detection. After that, the detection results of these sub-images are integrated to the original large-scene SAR image.

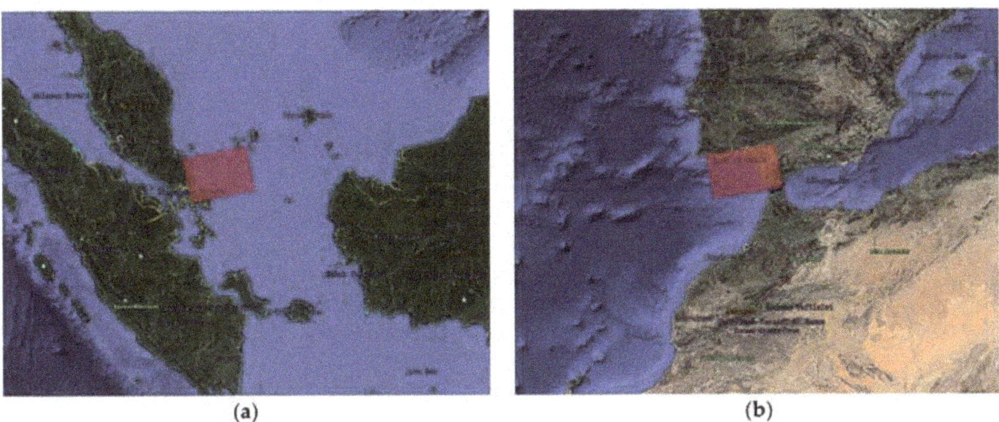

Figure 12. Coverage areas of two large-scene Sentinel-1 SAR images. (**a**) Singapore Strait; (**b**) Gulf of Cadiz.

Table 6. Descriptions of two large-scene Sentinel-1 SAR images.

No.	Place	Time	Polarization	Mode	Resolution (Range × Azimuth)	Image Size
Image 1	Singapore Strait	6 June 2020	VV	IW	5 m × 20 m	25,650 × 16,786
Image 2	Gulf of Cadiz	18 June 2020	VV	IW	5 m × 20 m	25,644 × 16,722

Figure 13 shows the visualization SAR ship detection results of Quad-FPN on the two large-scene SAR images. From Figure 13, most ships can be detected by Quad-FPN successfully, which shows its good migration application capability in ocean surveillance.

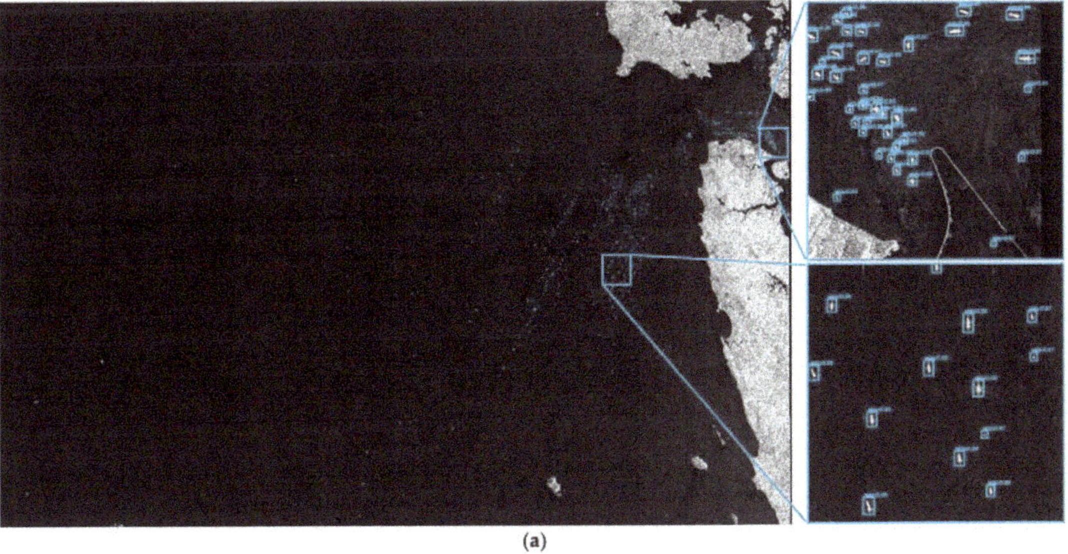

(**a**)

Figure 13. *Cont.*

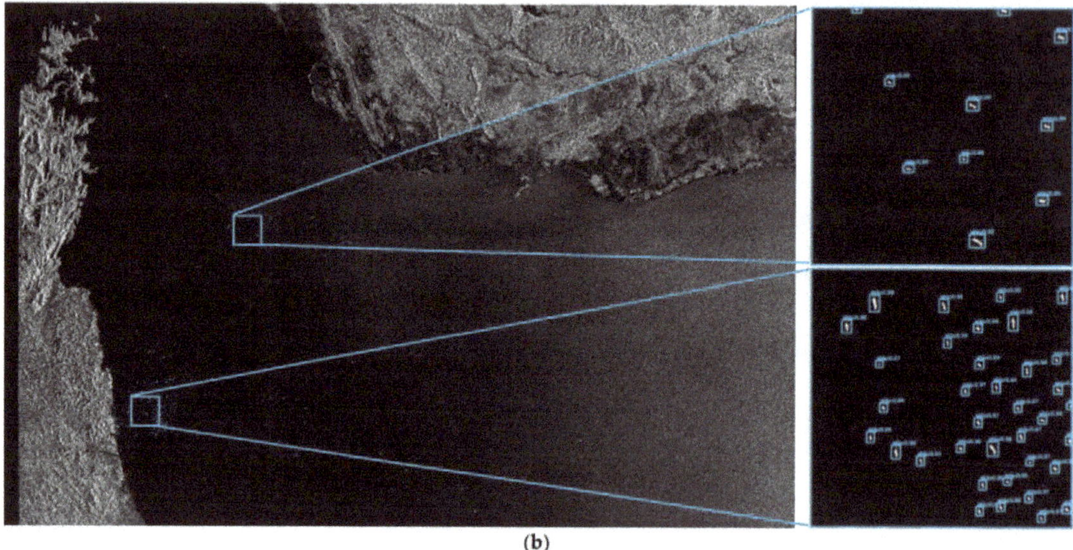

(b)

Figure 13. Detection results in two large-scene Sentinel-1 SAR images. (**a**) Image 1; (**b**) Image 2. Detections are marked by blue boxes.

4.3.1. Quantitative Comparison with State-of-The-Art

Tables 7 and 8 show their quantitative comparison with the other 12 competitive CNN-based SAR ship detectors. To be clear, in Tables 7 and 8, the GPU time is selected to compare their speed (t_{GPU}) because modern CNN-based detectors are always run on GPUs. From Tables 7 and 8, one can find that Quad-FPN achieves the best detection accuracy on the two large-scene SAR images, showing its good migration capability.

On the Image 1, Quad-FPN offers an accuracy of 83.96% mAP, superior to the second-best PANET [37] (83.96% mAP > 80.51% mAP); on the Image 2, Quad-FPN offers an accuracy of 87.03% mAP, superior to the second-best PANET [37] (87.03% mAP > 84.33% mAP). To be honest, we find that Quad-FPN's detection speed is relatively modest in contrast to others; thus, further detection speed improvements can be performed in the future.

4.3.2. Quantitative Comparison with CFAR

Finally, we perform an experiment to compare performance with a classical and common-used two-parameter CFAR detector. Following the standard implementation process from Deng et al. [44], we obtain the CFAR's detection results in the Sentinel-1 toolbox [45]. Tables 9 and 10 show their quantitative detection results.

Table 7. Quantitative evaluation indices comparison with the 12 state-of-the-art CNN-based SAR ship detectors on Image 1.

No.	Method	GT	Detections	TP	FP	FN	r (%)	p (%)	mAP (%)	t_{GPU} (s)
1	Faster R-CNN [22]	760	707	619	88	141	81.45	87.55	79.43	71.93
2	PANET [37]	760	725	630	95	130	82.89	86.90	<u>80.51</u>	76.33
3	Cascade R-CNN [21]	760	508	488	20	272	64.21	96.06	63.41	83.13
4	Double-Head R-CNN [38]	760	713	621	92	139	81.71	87.10	79.53	113.40
5	Grid R-CNN [39]	760	780	637	143	123	83.82	81.67	80.36	76.76
6	DCN [24]	760	715	618	97	142	81.32	86.43	79.33	74.36
7	Guided Anchoring [40]	760	451	434	17	326	57.11	96.23	55.93	102.20
8	Free-Anchor [41]	760	1187	651	536	109	85.66	54.84	79.55	67.09
9	HR-SDNet [12]	760	509	490	19	270	64.47	96.27	63.78	110.83
10	DAPN [13]	760	707	616	91	144	81.05	87.13	78.76	76.61
11	SER Faster R-CNN [14]	760	678	599	79	161	78.82	88.35	77.28	70.53
12	ARPN [15]	760	712	620	92	154	80.10	87.08	78.98	75.14
13	**Quad-FPN (Ours)**	760	904	662	242	98	87.11	73.23	**83.96**	122.97

The best detector is bold and the second-best is underlined.

Table 8. Quantitative evaluation indices comparison with the 12 state-of-the-art CNN-based SAR ship detectors on Image 2.

No.	Method	GT	Detections	TP	FP	FN	r (%)	p (%)	mAP (%)	t_{GPU} (s)
1	Faster R-CNN [22]	351	324	290	34	61	82.62	89.51	81.76	70.27
2	PANET [37]	351	337	299	38	52	85.19	88.72	<u>84.33</u>	74.93
3	Cascade R-CNN [21]	351	260	255	5	96	72.65	98.08	72.46	81.74
4	Double-Head R-CNN [38]	351	336	298	38	53	84.90	88.69	84.32	112.53
5	Grid R-CNN [39]	351	392	296	96	55	84.33	75.51	82.70	74.33
6	DCN [24]	351	334	291	43	60	82.91	87.13	81.52	73.00
7	Guided Anchoring [40]	351	222	219	3	132	62.39	98.65	62.21	101.37
8	Free-Anchor [41]	351	586	302	284	49	86.04	51.54	81.21	67.00
9	HR-SDNet [12]	351	264	256	8	95	72.93	96.97	72.51	106.88
10	DAPN [13]	351	335	292	43	59	83.19	87.16	82.38	75.25
11	SER Faster R-CNN [14]	351	323	291	32	60	82.91	90.09	81.94	70.56
12	ARPN [15]	351	348	298	50	62	82.78	85.63	81.09	76.28
13	**Quad-FPN (Ours)**	351	403	310	93	41	88.32	76.92	**87.03**	120.42

The best detector is bold and the second-best is underlined.

Table 9. Quantitative evaluation indices comparison with CFAR on Image 1.

Method	GT	Detections	TP	FP	FN	r (%)	p (%)	F1	t_{CPU} (s)
CFAR	760	863	603	260	157	79.34	69.87	0.74	884.00
Quad-FPN (Ours)	760	904	662	242	98	84.34	83.79	**0.84**	223.15

Table 10. Quantitative evaluation indices comparison with CFAR on Image 2.

Method	GT	Detections	TP	FP	FN	r (%)	p (%)	F1	t_{CPU} (s)
CFAR	351	556	314	242	37	89.46	56.47	0.69	735.00
Quad-FPN (Ours)	351	403	310	93	41	88.32	80.31	**0.84**	226.08

In Tables 9 and 10, the traditional CFAR usually does not use mAP from the DL community to measure accuracy, so F1 is used to represent accuracy, defined by:

$$F1 = 2 \times \frac{p \times r}{p + r} \tag{22}$$

Moreover, in Tables 9 and 10, CFARs are usually run on CPUs, whereas modern DL-based methods are always run on GPUs; to ensure a reasonable comparison, the CPU time is selected for their speed comparison (t_{CPU}). From Tables 9 and 10, Quad-FPN is greatly superior to CFAR in terms of the detection accuracy, i.e., 0.74 F1 of CFAR on Image 1 << 0.84 F1 of Quad-FPN on Image 1, and 0.69 F1 of CFAR on Image 2 << 0.84 F1 of Quad-FPN on Image 2. The detection speed of Quad-FPN is also greatly superior to CFAR, i.e., 223.15 s CPU time of Quad-FPN on Image 1 << 884.00 s CPU time of CFAR on Image 1, and 226.08 s CPU time of Quad-FPN on Image 2 << 735.00 s CPU time of CFAR on Image 2. Therefore, Quad-FPN might still meet the needs of practical applications.

5. Ablation Study

In this section, ablation studies are conducted to verify the effectiveness of each FPN. We also discuss the advantages of each innovation. Here, we take the SSDD dataset as an example to show the results, due to limited pages. Table 11 shows the effectiveness of the Quad-FPN pipeline (DE-CO-FPN→CA-FR-FPN→PA-SA-FPN→BS-GA-FPN). From Table 11, the detection accuracy is improved step by step from left to right in the Quad-FPN pipeline architecture (89.92% mAP→93.61% mAP→94.58% mAP→95.29% mAP). This can show each FPN's effectiveness from the perspective of the overall structure.

Table 11. Effectiveness of the Quad-FPN pipeline.

DE-CO-FPN	CA-FR-FPN	PA-SA-FPN	BS-GA-FPN	r (%)	p (%)	mAP (%)
✓	✗	✗	✗	91.18	82.12	89.92
✓	✓	✗	✗	94.30	84.38	93.61
✓	✓	✓	✗	95.04	86.89	94.58
✓	✓	✓	✓	95.77	89.52	**95.29**

To be clear, the sequence of the four FPNs is better kept unchanged; otherwise, the final accuracy cannot reach the best level according to our experiments. Some detailed analysis can be found in Section 2 (i.e., the overall design idea of Quad-FPN).

5.1. Ablation Study on DE-CO-FPN

We make two experiments with respect to DE-CO-FPN. Experiment 1 in Section 5.1.1 is used to confirm the effectiveness of DE-CO-FPN, directly. Experiment 2 in Section 5.1.2 is used to confirm the advantage of the deformable convolution.

5.1.1. Experiment 1: Effectiveness of DE-CO-FPN

Table 12 shows the ablation study results on DE-CO-FPM. In Table 12, "✗" denotes removing DE-CO-FPN (the other three FPNs are reserved) and "✓" denotes using DE-CO-FPN. From Table 12, DE-CO-FPN improves the accuracy by ~3% mAP, which shows its effectiveness. Combined with it, SAR ship features extracted by networks will contain useful shape information; moreover, they can alleviate complex background interferences.

Table 12. Effectiveness of DE-CO-FPN.

DE-CO-FPN	r (%)	p (%)	mAP (%)
✗	94.49	94.06	92.36
✓	95.77	89.52	**95.29**

5.1.2. Experiment 2: Different Types of Convolutions

Table 13 shows the ablation study results on different convolution types. In Table 13, "Standard" denotes the traditional regular convolution in Figure 3a, "Dilated" denotes the dilated convolution in Figure 3b, and "Deformable" denotes the deformable convolution in Figure 3c. From Table 13, the deformable convolution achieves the best detection accuracy because it can more effectively model various ships' shapes by its adaptive kernel offset learning. This adaptive kernel offset learning can extract the shape and edge features of ships accurately, to suppress the interference of complex backgrounds, especially for the complex inshore scenes. In this way, ships can be separated successfully from complex backgrounds. Thus, this deformable convolution process can be regarded as an extraction of salient objects in various scenes, which plays a role of spatial attention. Accordingly, the accuracy on the overall dataset is improved.

Table 13. Different types of convolutions.

Convolution Type	r (%)	p (%)	mAP (%)
Standard	94.49	94.06	92.36
Dilated	94.12	91.59	93.87
Deformable (Ours)	95.77	89.52	**95.29**

5.2. Ablation Study on CA-FR-FPN

With respect to CA-FR-FPN, we will make two experiments. Experiment 1 in Section 5.2.1 is used to confirm the effectiveness of CA-FR-FPN, directly. Experiment 2 in Section 5.2.2 is used to determine the appropriate feature amplification factor α in CA-FR-Module.

5.2.1. Experiment 1: Effectiveness of CA-FR-FPN

Table 14 shows the ablation study results on CA-FR-FPN. In Table 14, "✗" denotes removing CA-FR-FPN (i.e., not using the CA-FR-Module, but the other three FPNs are reserved.); "✓" denotes using the CA-FR-FPN. From Table 14, CA-FR-FPN improves the detection accuracy by ~1% mAP because it can be aware of more valuable information for feature up-sampling. Its adaptive content-aware kernel can improve the transmission benefits of information flow, to improve the detection performance. This is because it can effectively capture the rich semantic information required by dense detection tasks, especially for densely distributed small ships. This can avoid the feature loss because of small ship features' poor conspicuousness. Accordingly, the accuracy on the overall dataset is improved.

Table 14. Effectiveness of CA-FR-FPN.

CA-FR-FPN	r (%)	p (%)	mAP (%)
✗	95.22	74.78	94.74
✓	95.77	89.52	**95.29**

5.2.2. Experiment 2: Different Feature Amplification Factors

Table 15 shows the ablation study results on feature amplification factor α in CA-FR-Module. In Table 15, "✗" denotes not amplifying features. From Table 15, when features are amplified no matter what the value of α is, the detection accuracy can obtain improvements, compared with not amplifying features. Therefore, the feature amplification can indeed enhance the content-aware benefits of the kernel prediction, no matter what the value of α is. This is because in the embedded feature amplification space, the amount of information of feature maps will be effectively increased, promoting the better correctness of the kernel prediction. Finally, in our Quad-FPN, to obtain a better detection accuracy (95.29% mAP), α is set to an optimal or saturated value 8.

Table 15. Different feature amplification factors.

α	r (%)	p (%)	mAP (%)
✗	93.20	82.57	92.25
2	94.12	82.45	93.12
4	94.49	90.49	94.08
6	95.04	88.98	94.61
8	95.77	89.52	**95.29**
10	94.67	90.51	94.36
12	94.67	88.34	94.05
14	95.22	90.40	94.79
16	94.85	90.21	94.56
18	95.04	87.78	94.54

5.3. Ablation Study on PA-SA-FPN

We make three experiments with respect to PA-SA-FPN. Experiment 1 in Section 5.3.1 is used to confirm the effectiveness of PA-SA-FPN, directly. Experiment 2 in Section 5.3.2 is used to confirm the effectiveness of PA-SA-Module. Experiment 3 in Section 5.3.3 is used to confirm the advantage of PA-SA-Module.

5.3.1. Experiment 1: Effectiveness of PA-SA-FPN

Table 16 shows the ablation study results on PA-SA-FPN. In Table 16, "✗" denotes removing PA-SA-FPN (the other three FPNs are reserved); "✓" denotes using PA-SA-FPN. From Table 16, PA-SA-FPN improves the detection accuracy by ~1.5% mAP because the low-level spatial location information in the pyramid bottom has been transmitted to the top in PA-SA-FPN. In this way, the positionings of large ship bounding boxes will become more accurate. Accordingly, the accuracy on the overall dataset is improved.

Table 16. Effectiveness of PA-SA-FPN.

PA-SA-FPN	r (%)	p (%)	mAP (%)
✗	94.49	80.19	93.88
✓	95.77	89.52	**95.29**

5.3.2. Experiment 2: Effectiveness of PA-SA-Module

Table 17 shows the ablation study results on PA-SA-Module. From Table 17, PA-SA-Module can effectively enhance the detection accuracy by ~1% mAP because it can enable more pivotal spatial information in the pyramid bottom be effectively transmitted to the top. This can improve path aggregation benefits. In this way, the features of large ships might become richer and more discriminative. Accordingly, the accuracy on the overall dataset is improved.

Table 17. Effectiveness of PA-SA-Module.

PA-SA-Module	r (%)	p (%)	mAP (%)
✗	94.49	83.44	94.00
✓	95.77	89.52	**95.29**

5.3.3. Experiment 3: Different Attention Types

Table 18 shows the ablation study results on different attention types. In Table 18, "SE" denotes the squeeze-and-excitation mechanism [36] and "CBAM" denotes the convolutional block attention module [28]. From Table 18, PA-SA-Module is superior to others because it can cause key spatial global information to be transmitted more efficiently, which means that it is more suitable for PA-SA-FPN. Moreover, different from the previous CBAM, our designed space encoder $f_{space\text{-}encode}$ can encode the space information. It is can represent the spatial correlation more effectively. This can improve spatial attention gains because the features in the coding space are more concentrated.

Table 18. Different attention types.

Attention Type	r (%)	p (%)	mAP (%)
SE [36]	94.85	91.49	94.47
CBAM [28]	95.04	84.07	94.04
PA-SA-Module (Ours)	95.77	89.52	**95.29**

5.4. Ablation Study on BS-GA-FPN

We conduct three experiments with respect to BS-GA-FPN. Experiment 1 in Section 5.4.1 is used to confirm the effectiveness of BS-GA-FPN, directly. Experiment 2 in Section 5.4.2 is used to confirm the effectiveness of GA. Experiment 3 in Section 5.4.3 is used to confirm the advantage of GA.

5.4.1. Experiment 1: Effectiveness of BS-GA-FPN

Table 19 shows the ablation study results on BS-GA-FPN. In Table 19, "✗" denotes removing BS-GA-FPN (the other three FPNs are reserved); "✓" denotes using BS-GA-FPN. From Table 19, BS-GA-FPN can play an important role in ensuring higher detection accuracy because it can improve the accuracy by ~1% mAP. In this way, ship multi-scale features can be effectively balanced, which can achieve a stronger feature expression capacity of the final FPN. Accordingly, the accuracy on the overall dataset is improved.

Table 19. Effectiveness of BS-GA-FPN.

BS-GA-Module	r (%)	p (%)	mAP (%)
✗	95.04	86.89	94.58
✓	95.77	89.52	**95.29**

5.4.2. Experiment 2: Effectiveness of GA

Table 20 shows the ablation study results on GA. From Table 20, GA can improve the detection accuracy because when various ship multi-scale features are refined by it, they can become more discriminative. This feature self-attention might amplify important global information and suppress tiresome interferences, which can enhance the feature expressiveness of FPN. Essentially, GA is able to directly capture long-range dependence of each location (global response) through calculating the interaction between two different arbitrary positions. The whole GA refinement is essentially equivalent to construct a convolutional kernel with the same size as the feature map, to maintain more useful ship information. Accordingly, the accuracy on the overall dataset is improved.

Table 20. Effectiveness of GA.

GA	r (%)	p (%)	mAP (%)
✗	95.22	90.24	94.80
✓	95.77	89.52	**95.29**

5.4.3. Experiment 3: Different Refinement Types

Table 21 shows the ablation study results of different refinement types. In Table 21, we compare three refinement types, including a convolutional layer, an SE [36], and a CBAM [28]. From Table 21, GA offers the best detection accuracy because it can directly capture long-range dependence of each location (global response) to maintain more useful ship information that makes feature maps more discriminative. Different from the traditional convolution refinement types, its receptive field is wider, i.e., the whole input feature map's size, resulting in a better spatial correlation learning. Accordingly, the accuracy on the overall dataset is improved.

Table 21. Different refinement types.

Refinement Type	r (%)	p (%)	mAP (%)
Convolution	94.30	89.06	93.90
SE [36]	94.85	91.49	94.43
CBAM [28]	95.04	87.48	94.48
PA-SA-Module (Ours)	95.77	89.52	**95.29**

6. Conclusions

Aiming at some challenges in SAR ship detection, e.g., complex background interferences, multi-scale ship feature differences, and indistinctive small ship features, a novel Quad-FPN is proposed for SAR ship detection in this paper. Quad-FPN consists of four unique FPNs that can guarantee its excellent detection performance, i.e., DE-CO-FPN, CA-FR-FPN, PA-SA-FPN, and BS-GA-FPN. In DE-CO-FPN, we adopt the deformable convolution to extract SAR ship features that will contain more useful ship shape information, meanwhile alleviating complex background interferences. In CA-FR-FPN, we design a CA-FR-Module to enhance feature transmission benefits when performing the up-sampling multi-level feature fusion. In PA-SA-FPN, we add an extra path aggregation branch with a space attention module from the pyramid bottom to the top. In BS-GA-FPN, we further refine features from each feature level in the pyramid to address feature level imbalance of different scale ships. We perform extensive ablation studies to confirm the effectiveness of each FPN. Experimental results on five open datasets jointly reveal that Quad-FPN can offer the most superior SAR ship detection performance compared with the other 12 competitive state-of-the-art CNN-based SAR ship detectors. Moreover, the satisfactory detection results in two large-scene Sentinel-1 SAR images showing Quad-FPN's excellent migration capability in ocean surveillance. Quad-FPN is an excellent two-stage SAR ship detector. Four FPNs' internal implementations are different from previous work. They are well-designed improvements to ensure the state-of-the-art detection performance, without bells and whistles. They can exactly enable Quad-FPN's excellent ship scale-adaptability and detection scene-adaptability.

Our future work is as follows:

1. We will consider the cost of deformable convolutions, in the future.
2. We will consider optimizing the detection speed of Quad-FPN, in the future.
3. We will further study the effect of four FPNs' sequence on performance, in the future.

4. We will consider the challenges within SAR data, e.g., the azimuth ambiguity, side-lobes, and the sea state, to optimize Quad-FPN's detection performance, in the future.
5. We will consider making efforts to combine modern deep CNN abstract features and traditional concrete ones to further improve detection accuracy, in the future.

Author Contributions: Conceptualization, T.Z.; methodology, T.Z.; software, T.Z.; validation, T.Z.; formal analysis, T.Z.; investigation, T.Z.; resources, T.Z.; data curation, T.Z.; writing—original draft preparation, T.Z.; writing—review and editing, X.Z. and X.K.; visualization, T.Z.; supervision, X.Z.; project administration, X.Z.; funding acquisition, X.Z. All authors have read and agreed to the published version of the manuscript.

Funding: This work was supported by the National Natural Science Foundation of China (61571099).

Institutional Review Board Statement: Not applicable.

Informed Consent Statement: Not applicable.

Data Availability Statement: No new data were created or analyzed in this study. Data sharing is not applicable to this article.

Acknowledgments: The authors would like to thank the editors and the four anonymous reviewers for their valuable comments that can greatly improve our manuscript.

Conflicts of Interest: The authors declare no conflict of interest.

References

1. Liu, T.; Zhang, J.; Gao, G.; Yang, J.; Marino, A. CFAR Ship Detection in Polarimetric Synthetic Aperture Radar Images Based on Whitening Filter. *IEEE Trans. Geosci. Remote Sens.* **2020**, *58*, 58–81. [CrossRef]
2. Yang, M.; Guo, C.; Zhong, H.; Yin, H. A Curvature-Based Saliency Method for Ship Detection in SAR Images. *IEEE Geosci. Remote Sens. Lett.* **2020**, 1–5. [CrossRef]
3. Lin, H.; Chen, H.; Jin, K.; Zeng, L.; Yang, J. Ship Detection with Superpixel-Level Fisher Vector in High-Resolution SAR Images. *IEEE Geosci. Remote Sens. Lett.* **2020**, *17*, 247–251. [CrossRef]
4. Schwegmann, C.P.; Kleynhans, W.; Salmon, B.P. Synthetic Aperture Radar Ship Detection Using Haar-Like Features. *IEEE Geosci. Remote Sens. Lett.* **2017**, *14*, 154–158. [CrossRef]
5. Li, J.; Qu, C.; Shao, J. Ship detection in SAR images based on an improved faster R-CNN. In Proceedings of the SAR in Big Data Era: Models, Methods and Applications, Beijing, China, 13–14 November 2017; pp. 1–6.
6. Zhang, T.; Zhang, X.; Shi, J.; Wei, S. HyperLi-Net: A hyper-light deep learning network for high-accurate and high-speed ship detection from synthetic aperture radar imagery. *ISPRS J. Photogramm. Remote Sens.* **2020**, *167*, 123–153. [CrossRef]
7. Wang, Y.; Wang, C.; Zhang, H.; Dong, Y.; Wei, S. A SAR Dataset of Ship Detection for Deep Learning under Complex Backgrounds. *Remote Sens.* **2019**, *11*, 765. [CrossRef]
8. Wei, S.; Zeng, X.; Qu, Q.; Wang, M.; Su, H.; Shi, J. HRSID: A High-Resolution SAR Images Dataset for Ship Detection and Instance Segmentation. *IEEE Access* **2020**, *8*, 120234–120254. [CrossRef]
9. Li, J.; Qu, C.; Peng, S. A ship detection method based on cascade CNN in SAR images. *Control Decis.* **2019**, *34*, 2191–2197.
10. Zhang, T.; Zhang, X.; Shi, J.; Wei, S. Depthwise Separable Convolution Neural Network for High-Speed SAR Ship Detection. *Remote Sens.* **2019**, *11*, 2483. [CrossRef]
11. Yang, R.; Wang, G.; Pan, Z.; Lu, H.; Zhang, H.; Jia, X. A Novel False Alarm Suppression Method for CNN-Based SAR Ship Detector. *IEEE Geosci. Remote Sens. Lett.* **2020**, 1–5. [CrossRef]
12. Wei, S.; Su, H.; Ming, J.; Wang, C.; Yan, M.; Kumar, D.; Shi, J.; Zhang, X. Precise and Robust Ship Detection for High-Resolution SAR Imagery Based on HR-SDNet. *Remote Sens.* **2020**, *12*, 167. [CrossRef]
13. Cui, Z.; Li, Q.; Cao, Z.; Liu, N. Dense Attention Pyramid Networks for Multi-Scale Ship Detection in SAR Images. *IEEE Trans. Geosci. Remote Sens.* **2019**, *57*, 8983–8997. [CrossRef]
14. Lin, Z.; Ji, K.; Leng, X.; Kuang, G. Squeeze and Excitation Rank Faster R-CNN for Ship Detection in SAR Images. *IEEE Geosci. Remote Sens. Lett.* **2019**, *16*, 751–755. [CrossRef]
15. Zhao, Y.; Zhao, L.; Xiong, B.; Kuang, G. Attention Receptive Pyramid Network for Ship Detection in SAR Images. *IEEE J. Sel. Top. Appl. Earth Obs. Remote Sens.* **2020**, *13*, 2738–2756. [CrossRef]
16. Girshick, R. Fast R-CNN. In Proceedings of the IEEE International Conference on Computer Vision (ICCV), Santiago, Chile, 7–13 December 2015; pp. 1440–1448.
17. Ren, S.; He, K.; Girshick, R.; Sun, J. Faster R-CNN: Towards Real-Time Object Detection with Region Proposal Networks. *IEEE Trans. Pattern Anal. Mach. Intell.* **2017**, *39*, 1137–1149. [CrossRef]
18. Redmon, J.; Farhadi, A. YOLOv3: An Incremental Improvement. *arXiv* **2018**, arXiv:1804.02767.

19. Lin, T.-Y.; Goyal, P.; Girshick, R.; He, K.; Dollar, P. Focal Loss for Dense Object Detection. *IEEE Trans. Pattern Anal. Mach. Intell.* **2020**, *42*, 318–327. [CrossRef]
20. Liu, W.; Anguelov, D.; Erhan, D.; Szegedy, C.; Reed, S.; Fu, C.-Y.; Berg, A.C. SSD: Single Shot MultiBox Detector. In Proceedings of the European Conference on Computer Vision (ECCV), Amsterdam, The Netherlands, 8–16 October 2016; Springer International Publishing: Cham, Switzerland, 2016; pp. 21–37.
21. Cai, Z.; Vasconcelos, N. Cascade R-CNN: Delving into High Quality Object Detection. In Proceedings of the IEEE/CVF Conference on Computer Vision and Pattern Recognition (CVPR), Salt Lake City, UT, USA, 18–23 June 2018; pp. 6154–6162.
22. Lin, T.-Y.; Dollar, P.; Girshick, R.; He, K.; Hariharan, B.; Belongie, S. Feature Pyramid Networks for Object Detection. In Proceedings of the IEEE Conference on Computer Vision and Pattern Recognition (CVPR), Honolulu, HI, USA, 21–26 July 2017; pp. 936–944.
23. Github. Available online: https://github.com/TianwenZhang0825/Quad-FPN (accessed on 25 June 2021).
24. Dai, J.; Qi, H.; Xiong, Y.; Li, Y.; Zhang, G.; Hu, H.; Wei, Y. Deformable Convolutional Networks. In Proceedings of the IEEE International Conference on Computer Vision (ICCV), Venice, Italy, 22–29 October 2017; pp. 764–773.
25. Yu, F.; Koltun, V. Multi-scale context aggregation by dilated convolutions. *arXiv* **2016**, arXiv:1511.07122.
26. Wang, J.; Chen, K.; Xu, R.; Liu, Z.; Loy, C.C.; Lin, D. CARAFE: Content-Aware ReAssembly of FEatures. In Proceedings of the IEEE/CVF International Conference on Computer Vision (ICCV), Seoul, Korea, 27 October–2 November 2019; pp. 3007–3016.
27. Shi, W.; Caballero, J.; Huszar, F.; Totz, J.; Aitken, A.P.; Bishop, R.; Rueckert, D.; Wang, Z. Real-Time Single Image and Video Super-Resolution Using an Efficient Sub-Pixel Convolutional Neural Network. In Proceedings of the IEEE Conference on Computer Vision and Pattern Recognition (CVPR), Las Vegas, NV, USA, 27–30 June 2016; pp. 1874–1883.
28. Woo, S.; Park, J.; Lee, J.-Y.; Kweon, I.S. CBAM: Convolutional Block Attention Module. *arXiv* **2018**, arXiv:1807.06521.
29. Lin, M.; Chen, Q.; Yan, S. Network in Network. *arXiv* **2013**, arXiv:1312.4400.
30. Pang, J.; Chen, K.; Shi, J.; Feng, H.; Ouyang, W.; Lin, D. Libra R-CNN: Towards Balanced Learning for Object Detection. *arXiv* **2019**, arXiv:1904.02701.
31. Wang, X.; Girshick, R.; Gupta, A.; He, K. Non-local Neural Networks. *arXiv* **2017**, arXiv:1711.07971.
32. Chen, K.; Wang, J.; Pang, J.; Cao, Y.; Xiong, Y.; Li, X.; Sun, S.; Feng, W.; Liu, Z.; Xu, J.; et al. MMDetection: Open MMLab Detection Toolbox and Benchmark. *arXiv* **2019**, arXiv:1906.07155.
33. He, K.; Girshick, R.; Doll´ar, P. Rethinking ImageNet Pre-Training. *arXiv* **2019**, arXiv:1811.08883.
34. Goyal, P.; Dollár, P.; Girshick, R.; Noordhuis, P.; Wesolowski, L.; Kyrola, A.; Tulloch, A.; Jia, Y.; He, K. Accurate, Large Minibatch SGD: Training ImageNet in 1 Hour. *arXiv* **2017**, arXiv:1706.02677.
35. Bodla, N.; Singh, B.; Chellappa, R.; Davis, L.S. Soft-NMS—Improving Object Detection with One Line of Code. In Proceedings of the IEEE International Conference on Computer Vision (ICCV), Venice, Italy, 22–29 October 2017; pp. 5562–5570.
36. Hu, J.; Shen, L.; Albanie, S.; Sun, G.; Wu, E. Squeeze-and-Excitation Networks. *IEEE Trans. Pattern Anal. Mach. Intell.* **2020**, *42*, 2011–2023. [CrossRef]
37. Liu, S.; Qi, L.; Qin, H.; Shi, J.; Jia, J. Path Aggregation Network for Instance Segmentation. In Proceedings of the IEEE/CVF Conference on Computer Vision and Pattern Recognition (CVPR), Salt Lake City, UT, USA, 18–23 June 2018; pp. 8759–8768.
38. Wu, Y.; Chen, Y.; Yuan, L.; Liu, Z.; Wang, L.; Li, H.; Fu, Y. Rethinking Classification and Localization for Object Detection. In Proceedings of the IEEE/CVF Conference on Computer Vision and Pattern Recognition (CVPR), Seattle, DC, USA, 16–18 June 2020; pp. 10183–10192.
39. Lu, X.; Li, B.; Yue, Y.; Li, Q.; Yan, J. Grid R-CNN. In Proceedings of the 2019 IEEE/CVF Conference on Computer Vision and Pattern Recognition (CVPR), Long Beach, CA, USA, 16–20 June 2019; pp. 7355–7364.
40. Wang, J.; Chen, K.; Yang, S.; Loy, C.C.; Lin, D. Region Proposal by Guided Anchoring. In Proceedings of the 2019 IEEE/CVF Conference on Computer Vision and Pattern Recognition (CVPR), Long Beach, CA, USA, 16–20 June 2019; pp. 2960–2969.
41. Zhang, X.; Wan, F.; Liu, C.; Ji, R.; Ye, Q. FreeAnchor: Learning to Match Anchors for Visual Object Detection. In Proceedings of the International Conference on Neural Information Processing Systems (NIPS), Vancouver, CO, Canada, 10–12 December 2019; pp. 147–155.
42. Torres, R.; Snoeij, P.; Geudtner, D.; Bibby, D.; Davidson, M.; Attema, E.; Potin, P.; Rommen, B.; Floury, N.; Brown, M.; et al. GMES Sentinel-1 mission. *Remote Sens. Environ.* **2012**, *120*, 9–24. [CrossRef]
43. Zhang, T.; Zhang, X.; Ke, X.; Zhan, X.; Shi, J.; Wei, S.; Pan, D.; Li, J.; Su, H.; Zhou, Y.; et al. LS-SSDD-v1.0: A Deep Learning Dataset Dedicated to Small Ship Detection from Large-Scale Sentinel-1 SAR Images. *Remote Sens.* **2020**, *12*, 2997. [CrossRef]
44. Deng, Z.; Sun, H.; Zhou, S.; Zhao, J.; Lei, L.; Zou, H. Multi-scale object detection in remote sensing imagery with convolutional neural networks. *ISPRS J. Photogramm. Remote Sens.* **2018**, *145*, 3–22. [CrossRef]
45. Sentinel-1 Toolbox. Available online: https://sentinels.copernicus.eu/web/ (accessed on 4 April 2021).

 *remote sensing*

MDPI

Article

ADT-Det: Adaptive Dynamic Refined Single-Stage Transformer Detector for Arbitrary-Oriented Object Detection in Satellite Optical Imagery

Yongbin Zheng *,†**, Peng Sun** †**, Zongtan Zhou, Wanying Xu and Qiang Ren**

College of Intelligence Science and Technology, National University of Defense Technology, Changsha 410073, China; sunpeng@nudt.edu.cn (P.S.); ztzhou@nudt.edu.cn (Z.Z.); wanyingxu@nudt.edu.cn (W.X.); renqiang@nudt.edu.cn (Q.R.)
* Correspondence: zybnudt@nudt.edu.cn
† Y. Zheng and P. Sun contributed equally to this work.

Citation: Zheng, Y.; Sun, P.; Zhou, Z.; Xu, W.; Ren, Q. ADT-Det: Adaptive Dynamic Refined Single-Stage Transformer Detector for Arbitrary-Oriented Object Detection in Satellite Optical Imagery. *Remote Sens.* **2021**, *13*, 2623. https://doi.org/10.3390/rs13132623

Academic Editors: Anwaar Ulhaq, Douglas Pinto Sampaio Gomes and Danfeng Hong

Received: 19 May 2021
Accepted: 30 June 2021
Published: 4 July 2021

Publisher's Note: MDPI stays neutral with regard to jurisdictional claims in published maps and institutional affiliations.

Abstract: The detection of arbitrary-oriented and multi-scale objects in satellite optical imagery is an important task in remote sensing and computer vision. Despite significant research efforts, such detection remains largely unsolved due to the diversity of patterns in orientation, scale, aspect ratio, and visual appearance; the dense distribution of objects; and extreme imbalances in categories. In this paper, we propose an adaptive dynamic refined single-stage transformer detector to address the aforementioned challenges, aiming to achieve high recall and speed. Our detector realizes rotated object detection with RetinaNet as the baseline. Firstly, we propose a feature pyramid transformer (FPT) to enhance feature extraction of the rotated object detection framework through a feature interaction mechanism. This is beneficial for the detection of objects with diverse patterns in terms of scale, aspect ratio, visual appearance, and dense distributions. Secondly, we design two special post-processing steps for rotated objects with arbitrary orientations, large aspect ratios and dense distributions. The output features of FPT are fed into post-processing steps. In the first step, it performs the preliminary regression of locations and angle anchors for the refinement step. In the refinement step, it performs adaptive feature refinement first and then gives the final object detection result precisely. The main architecture of the refinement step is dynamic feature refinement (DFR), which is proposed to adaptively adjust the feature map and reconstruct a new feature map for arbitrary-oriented object detection to alleviate the mismatches between rotated bounding boxes and axis-aligned receptive fields. Thirdly, the focus loss is adopted to deal with the category imbalance problem. Experiments on two challenging satellite optical imagery public datasets, DOTA and HRSC2016, demonstrate that the proposed ADT-Det detector achieves a state-of-the-art detection accuracy (79.95% mAP for DOTA and 93.47% mAP for HRSC2016) while running very fast (14.6 fps with a 600 × 600 input image size).

Keywords: arbitrary-oriented object detection in satellite optical imagery; adaptive dynamic refined single-stage transformer detector; feature pyramid transformer; dynamic feature refinement

1. Introduction

In the past few decades, Earth observation satellites have been monitoring changes in the Earth's surface and the amount and resolution of satellite optical images have been greatly improved. The task of object detection in satellite optical images is to localize interest objects (such as vehicles, ships, aircraft, buildings, airports, ports) and identify their categories. This has numerous practical applications in satellite remote sensing and computer vision, warning of natural disasters, Earth surveying and mapping, and surveillance and traffic planning. Much progress in general-purpose horizontal detectors has been achieved by advances in deep convolutional neural networks (DCNNs) and the emergence of large datasets [1]. However, unlike natural images that are usually taken from horizontal perspectives, satellite optical images are taken with a bird's eye view, which often leads to the arbitrary orientation of objects in satellite images [2], as shown in Figure 1. Moreover, as mentioned in [2–4], the following significant challenges further increase the difficulty of object detection in satellite optical images:

- Large-scale difference. Objects in satellite images vary in size hugely [5]. There are small objects such as cars, ships, aircraft, and small houses in satellite images, as well as large objects such as ports, airports, ground track fields, bridges, and large buildings. In addition, the size of objects within the same category (such as large aircraft and small aircraft) in the same image also varies greatly.
- Dense distribution. There are many densely distributed objects in satellite optical images, such as cars and ships [5].
- Large aspect ratio. There are lots of objects with large aspect ratios, such as large vehicles, ships, harbors, and bridges in satellite optical images. The mismatch between the ground truth bounding box and the predicted bounding box of these objects is very sensitive to the rotation angle of objects [4].
- Category imbalance. Satellite optical imagery datasets are long-tailed, and the number of instances in each category varies greatly. For example, the amount of small vehicles is about 105 times larger than that of soccer ball fields in satellite optical imagery.

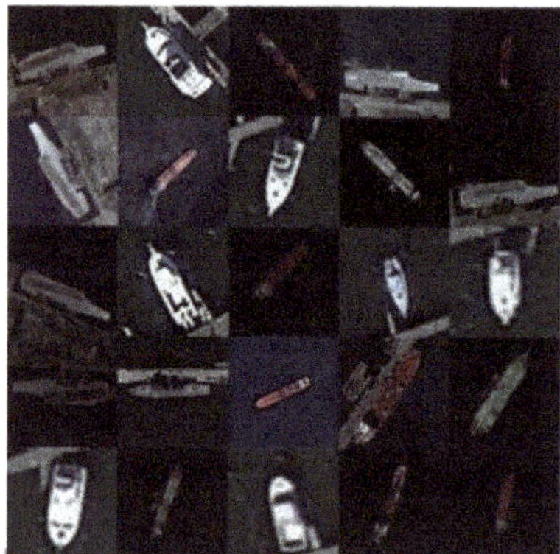

Figure 1. Examples of objects with various orientations in satellite optical imagery.

Recent research [6–9] has focused on the design of rotation detectors, which apply rotated regions of interest (RRoI) instead of horizontal regions of interest (HRoI). To meet the above challenges, a framework for rotated object detection consisting of a rotation learning stage and a feature refinement stage is proposed to improve the detection accuracy. Despite the fact that some newly developed rotated object detection methods [10–14] have made some progress in this area, their performance still falls considerably below that required for real-world applications. A main reason for their low detection performance is improper feature extraction for instances with arbitrary orientations, large aspect ratios, and dense distributions. As shown in Figure 2a, the general receptive field of deep neural network-based detectors is axis-aligned and square, representing a mismatch with the actual shape of the instances, and this usually produces false detections. Thus, our goal is to design a special feature pyramid transformer and feature refinement module which can be adjusted adaptively according to the angle and scale of the instance, as shown in Figure 2b. Then, we introduce the above methods into the rotated object detection framework to help extract more accurate features.

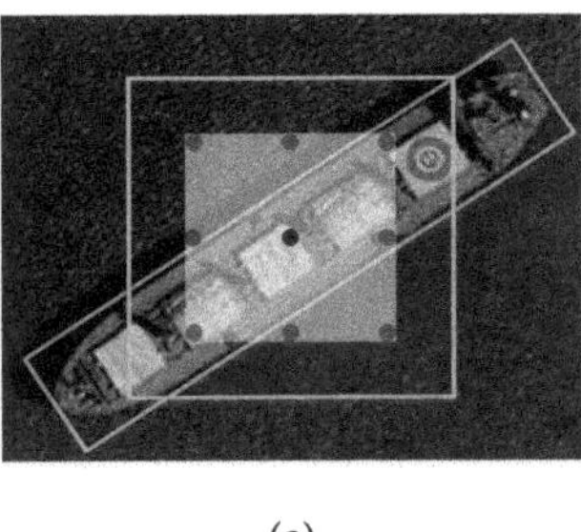 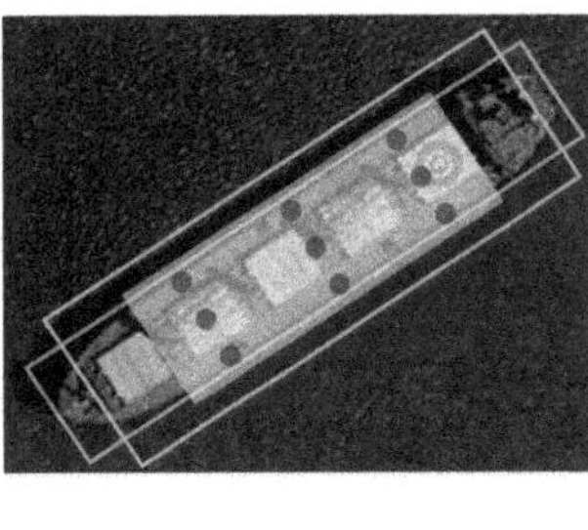

(a)   (b)

Figure 2. Comparison of receptive fields between (**a**) an axis-aligned neuron and (**b**) an adaptive neuron. The green rectangle represents the boundary of the instance, and the gray rectangle represents the boundary of the receptive field.

In this paper, we propose an adaptive dynamic refined single-stage transformer detector to address the aforementioned challenges, aiming to achieve a high recall and speed. Our detector realizes rotated object detection with RetinaNet as the baseline. Firstly, the feature pyramid transformer (FPT) is introduced into the traditional feature pyramid network (FPN) to enhance feature extraction through a feature interaction mechanism. This is beneficial for the detection of multi-scale objects and densely distributed objects. Secondly, the output features of FPT are fed into two post-processing steps. In the first step, the preliminary regression of locations and angle anchors for the refinement step is performed. In the refinement step, adaptive feature refinement is performed first and then the final object detection result is given precisely. The main architecture of the refinement step is the dynamic feature refinement (DFR), which is proposed to adaptively adjust the feature map and reconstruct a new feature map for arbitrary-oriented object detection to alleviate the mismatches between rotated bounding boxes and axis-aligned receptive fields. Experiments are carried out on two challenging satellite optical imagery public datasets, DOTA and HRSC2016, to demonstrate that our method outperforms previous state-of-the-art methods while running very fast.

The contributions of this work are three-fold:

(1) We propose a feature pyramid transformer for the feature extraction of the rotated object detection framework. This is beneficial for detecting objects with diverse patterns in terms of scale, aspect ratio, and visual appearance, and helps with the handling of challenging scenes with densely distributed instances through a feature interaction mechanism.

(2) We propose a dynamic feature refinement method for rotated objects with arbitrary orientations, large aspect ratios, and dense distributions. This can help to alleviate the bounding box mismatch problem.

(3) The proposed ADT-Det detector outperforms previous state-of-the-art detectors in terms of accuracy while running very fast.

2. Related Studies

Along with the wide application of satellite remote sensing and unmanned aerial vehicles, the amount of satellite optical imagery is increasing tremendously and object detection in satellite optical imagery has received increasing attention in the computer vision and remote sensing communities. Researchers have introduced DCNN-based detectors for object detection in satellite optical imagery, and oriented bounding boxes have been used instead of horizontal bounding boxes to reduce the mismatch between the predicted bounding box and corresponding objects. DCNN-based detectors are now reported as state-of-the-art.

In this section, we briefly review some previous well-known object detection methods in satellite or aerial optical images. In Section 2.1, we review the current mainstream detectors used for satellite optical image detection. In Section 2.2, we summarize some classical designs of DCNN-based detectors that can improve the detection performance.

2.1. The Mainstream Detectors for Object Detection in Satellite Optical Imagery

The current mainstream detectors for satellite optical image detection are rotation detectors. Existing rotation detectors are mostly employed as alternatives to horizontal bounding boxes. Generally, these detectors can be organized into two main categories: multi-stage detectors and single-stage detectors.

The framework of multi-stage detectors includes a pre-processing stage for region proposal and one or more post-processing stages to regress the bounding box of an object and identify its category. In the pre-processing stage, classification-independent region proposals are generated from an input image. Then, CNNs with a special architecture are used to subsequently extract features from these regions, and regression and classification are performed over the next several stages [3,4]. In the last stage, the final detection results are generated by non-maximum suppression (NMS) or other methods. To the best of our knowledge, RoI-Transformer [2] and SCRDet [15] are state-of-the-art multi-stage rotated objects detectors. The RoI-Transformer is a two-stage rotated object detector. Its first stage is a RRoI Learner that generates a transformation from a horizontal bounding box to an oriented bounding box by learning from the annotated data. One important task in the second stage is RoI alignment, which extracts rotation-invariant features from the oriented RoI for subsequent object regression and classification. SCRDet introduced SF-Net [16] and MDA-Net into Faster-RCNN [17] to detect small and densely distributed objects. By introducing the Intersection over Union (IoU) factor into the traditional smooth L_1 loss function, the IoU-Smooth L_1 Loss enables the angle regression to be more concise. Generally, the numerous redundant region proposals make multi-stage detectors more accurate than anchor-free detectors. However, they rely on a more complicated structure, which greatly reduces their speed.

Single-stage object detectors drop the complex and redundant region proposal network, directly regress the bounding box, and identify the category of objects. YOLO [18–20] treats object detection as a regression task. Image pixels are regressed to spatially separate bounding boxes and associate them with class probabilities using the GoogLeNet network. Its improved versions are YOLOv2 and YOLO9000, in which GoogLeNet is replaced by a simpler Dark-Net19 and some special strategies (e.g., batch normalization) are introduced. Liu et al. [21] proposed SSD to preserve the real-time speed while keeping the detection accuracy as high as possible. Just like YOLO, a fixed number of bounding boxes and scores are predicted for the presence of object category in these boxes, followed by a NMS [22] step to generate the final detection result. As observed in [5], the detection performance of general single-stage methods is considerably lower than that of multistage methods. Recently, R^{3}Det [4] and R^{4}Det [3] demonstrated high performance in detecting rotated objects in satellite optical images. R^{3}Det adopts RetinaNet [23] for the baseline and adds refinement to the network. The focal loss alleviates any imbalance between positive and negative samples. R^{4}Det proposed a single-stage object detection framework by introducing the recursive feature pyramid (RFP) into RetinaNet to integrate feature maps of different levels.

2.2. General Designs for DCNN-Based Object Detection in Satellite Optical Imagery

2.2.1. Feature Pyramid Networks (FPN)

In many DCNN-based object detection frameworks, FPN is a basic component used to extract multi-level features for detecting objects at different scales. Low-level features represent less semantic information but the resolution is higher; on the contrary, high-level features represent more semantic information but the resolution is lower. In order to make full use of low-level features and high-level features at the same time, Lin et al. [24] proposed a generic FPN approach to fuse a multi-scale feature pyramid with a top-down pathway and lateral connections. This has become the benchmark and performs well in feature extraction. Using a feature pyramid transformer [25] is an effective way to perform feature interaction between different scales and spaces. The transformed feature pyramid has a richer context than the original pyramid while maintaining the same size. In this paper, we introduce an FPT to enhance feature interaction in the feature fusion step.

2.2.2. Spatial Transformer Network

Atrous convolution [26] is an initial spatial transformer network. It increases the reception field by injecting holes into the standard convolution. Many improvements in dilated convolution have been proposed in recent years. Atrous spatial pyramid pooling (ASPP) [27] and denseASPP [28] obtained better results by cascading convolutions with different dilated rates in various forms. The Deformable Convolutional Network (DCN) [29] provides new ideas for spatial transformer networks. DCN can adjust the convolution kernels to make the receptive field more suitable for the feature map. General convolution is mostly horizontal and square. DCN can dynamically adjust according to the feature shape. We expect that it can improve the detection performance by introducing DCN into the feature extraction for rotated object detection.

2.2.3. Refined Object Detectiors

The research in [30] indicates that a low IoU threshold usually produces noisy detections. However, due to the mismatch between the optimal IoU of the detector and the IoU of the input hypothesis, detection performance tends to degrade as the IoU thresholds increase. To address these problems, Cascade RCNN [30] uses multiple stages with sequentially increasing IoU thresholds to train detectors. The main idea of RefineDet [31] is to coarsely adjust the locations and sizes of anchors using an anchor refinement module first. This is then followed by a regression branch to obtain more precise box information. Unlike two-stage detectors, the currently single-stage detector with a refinement stage is not well resolved in this respect. Feature misalignment is still one of the main reasons for the poor performance of refined single-stage detectors.

In this paper, we propose an adaptive dynamic refined single-stage transformer detector to address the aforementioned challenges, aiming to achieve a high recall and speed. Our detector realizes rotated object detection with RetinaNet as the baseline to achieve the detection of multi-scale objects and densely distributed objects. Firstly, the feature pyramid transformer (FPT) is introduced into the traditional feature pyramid network (FPN) to enhance feature extraction through a feature interaction mechanism. Secondly, the output features of FPT are fed into two post-processing steps considering the mismatch between the rotated bounding box and the general axis-aligned receptive fields of CNN. Dynamic Feature Refinement (DFR) is introduced to the refinement step. The key idea of DFR is to adaptively adjust the feature map and reconstruct a new feature map for arbitrary-oriented object detection to alleviate the mismatches between the rotated bounding box and the axis-aligned receptive fields. Extensive experiments and ablation studies show that our method can achieve state-of-the-art results in the task of object detection.

3. Methodology

In this section, we first describe our network architecture for arbitrary rotated object detection in Section 3.1. We then propose the feature pyramid transformer and dynamic feature refinement, which are our main contributions, in Sections 3.2 and 3.3, respectively. Finally, we show the details of our RetinaNet-based rotation detection method and the loss function in Section 3.4.

3.1. Network Architecture

The overall architecture of the proposed ADT-Det detector is sketched in Figure 3. Our pipeline improves upon RetinaNet and consists of a backbone network and two post-processing steps. The FPN network is utilized as the backbone and a feature pyramid transformer is proposed to enhance feature extraction for densely distributed instances.

Then, the backbone is attached in the post-processing steps. These consist of two sub-steps: first, a sub-step and a refinement sub-step, which will be described in detail in Sections 3.3 and 3.4. In the first sub-step, the preliminary regression of locations and angle anchors for the refinement sub-step is performed. In the refinement sub-step, adaptive feature refinement is performed first and then the final object detection result is given precisely. The main architecture of the refinement sub-step is the dynamic feature refinement (DFR), which is proposed to adaptively adjust the feature map and reconstruct a new feature map for rotated object detection (the detailed architecture of DFR is shown in Section 3.3). In the refinement sub-step, the feature fusion module (FFM) is considered as an important step to dynamically counteract the mismatch between the rotating object and the axis-aligned receptive fields of neurons. The overall framework is end-to-end trainable with a high efficiency.

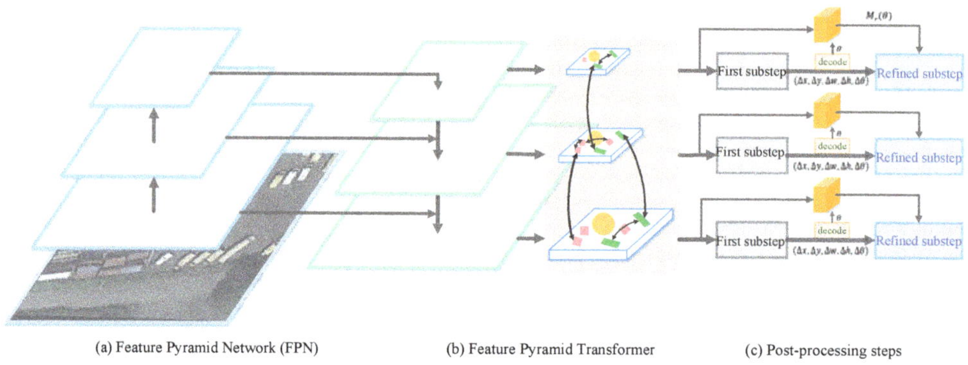

Figure 3. The framework of the proposed ADT-Det detector. Our pipeline consists of a backbone network and two post-processing steps. An FPN network is used as backbone network and a feature pyramid transformer is proposed to enhance the feature extraction. Then, the backbone is attached in the post-processing steps, which consist of two sub-steps: first, a sub-step and a refinement step. In the first sub-step, the preliminary regression of locations and angles for the refinement sub-step is performed. In the refinement sub-step, adaptive feature refinement is performed first and then the final object detection result is given precisely.

3.2. Feature Pyramid Transformer

We introduce a feature pyramid transformer (FPT) and add it between the backbone FPN network and the post-processing network to produce features with stronger semantic information. Its architecture is shown in Figure 4. Firstly, the features from FPN are transformed and re-arranged. Then, the output features are concatenated with the original feature map to obtain the concatenated features. Finally, the Conv3×3 operation is carried out to reduce the channel and obtain the transformed feature pyramid.

The FPT is a light network that enhances features through feature interaction with multiple scales and layers. It allows features of different levels to interact across space and scale. The FPT consists of three transformer steps: a self-transformer, a grounding transformer, and a rendering transformer. The self-transformer is introduced to capture objects that appear simultaneously on the same feature map. The grounding transformer is a up-bottom non-local interaction transformer that is used to enhance shallow features with different levels of features. As shown in Figure 5a,b, the inputs of the self-transformer and the grounding transformer are q_i, k_j, and v_j, where $q_i = f_q(X_i)$ represents the i-th query; $k_j = f_k(X_j)$ represents the j-th key; $v_j = f_v(X_j)$ represents the j-th value; and $f_q(.)$, $f_k(.)$, and $f_v(.)$ are used to perform queries, keys, and values operations on the feature map, respectively. The self-transformer adopts dot products as similarity function F_{sim} to capture co-occurring features in the same feature map. The output of F_{sim} is fed to the normalization function F_{norm} to generate weights $w_{(i,j)}$. Lastly, we multiply v_j and $w_{(i,j)}$ to obtain the transformed feature X. Unlike the self-transformer, the grounding transformer is a top-down non-local interaction that is used to strengthen shallow features with deep features. It uses Euclidean distance to measure the similarity of deep features and shallow features. The rendering transformer works with a bottom-up transformer to interact with the entire feature map, presenting higher-level semantic features in lower-level features. The transformation process is shown in Figure 5c. First, we calculate the weight w of Q through global average pooling from the shallow feature K. Then, the weights of Q (Q_{att}) and V are refined by Conv3×3 to reduce the size of the feature map. Finally, the refined Q_{att} and down-sampled V (V_{down}) are summed and processed by another Conv3×3 for rendering.

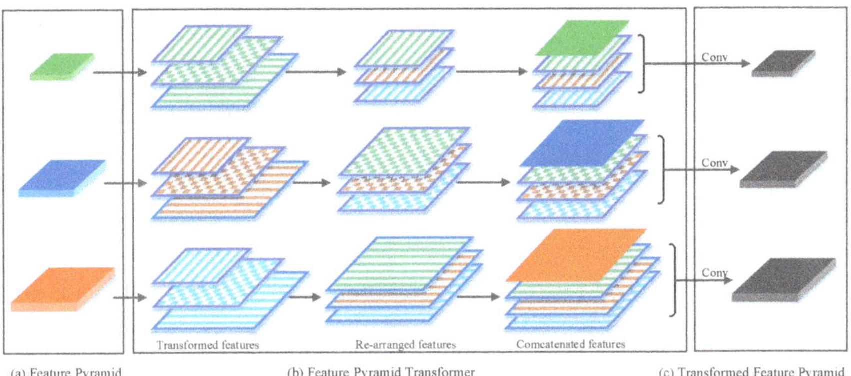

Figure 4. Three transformer steps: (**a**) self-transformer, (**b**) grounding transformer, (**c**) rendering transformer. $q_i = f_q(X_i)$ represents the i-th query, $k_j = f_k(X_j)$ represents the j-th key, and $v_j = f_v(X_j)$ represents the j-th value, where $f_q(.)$, $f_k(.)$, and $f_v(.)$ are used to perform queries, keys, and values operations on the feature map, respectively.

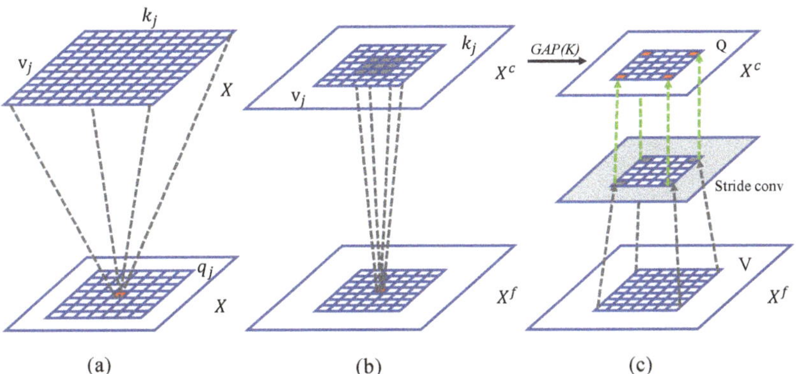

Figure 5. Architecture of the proposed feature pyramid transformer: (**a**) self-transformer, (**b**) grounding transformer, (**c**) rendering transformer. Firstly, the features from FPN are transformed and re-arranged. Then, the output features are concatenated with the original feature map to obtain the concatenated features. Finally, the Conv3×3 operation is carried out to reduce the channel and obtain the transformed feature pyramid.

3.3. Dynamic Feature Refinement

When detecting instances with arbitrary orientations, large aspect ratios, and dense distributions, the main reason for low detection performance is the feature misalignment problem, which is caused by differences in the scale and rotation between the orientated bounding box and the axis-aligned receptive fields. To alleviate the feature misalignment problem, we introduce dynamic feature refinement (DFR) to obtain the refined accurate bounding box. The architecture of DFR is shown in the bottom of Figure 6.

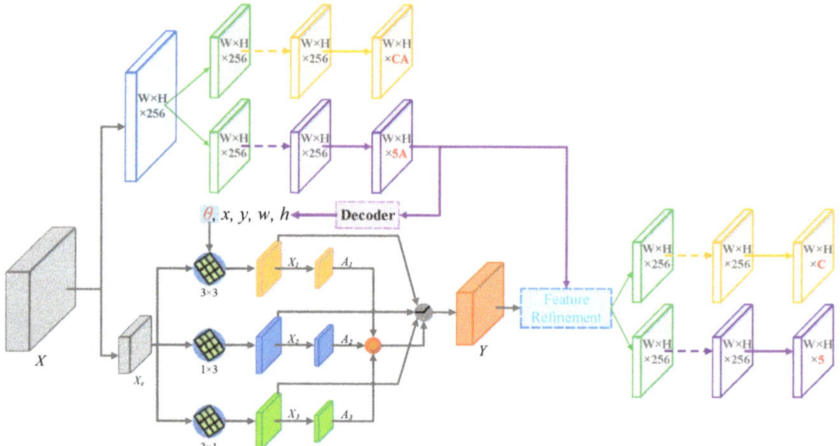

Figure 6. Architecture of the post-processing step. This consists of two sub-steps: the first sub-step and the refinement sub-step. Top: the first sub-step, which performs the preliminary regression of angle anchors for the refinement sub-step. Bottom: the refinement sub-step, which performs feature fusion and adaptive feature refinement and then gives the final object detection result precisely. On the left of the refinement sub-step is the feature fusion module, followed by the feature refinement module. On the right are two subnetworks, which perform object classification and regression.

We adopt a feature fusion module (FFM) to counteract the mismatches between arbitrary-orientation objects and axis-aligned receptive fields. This can dynamically and adaptively aggregate the features extracted by various kernel sizes, shapes (aspect ratios), and angles. The FFM takes the i-th stage feature map $X \in \mathbb{R}^{H \times W \times C}$ as an input and consists of two branches. In one branch, $X \in \mathbb{R}^{H \times W \times C}$ is connected to the classification and regression subnetworks to decode the location feature information. This is a normal network introduced from RetinaNet. The task of this branch is to generate initial location information and decode the angle feature information. In the other branch, we compress $X \in \mathbb{R}^{H \times W \times C}$ with a Conv1×1 layer and aggregate the improved information using batch normalization and ReLU. In order to further deal with the mismatches between rotated objects and axis-aligned receptive fields, we introduce the adaptive convolution (AdaptConv) into our DFR.

The AdaptConv is inspired by [32], and the implementation details are illustrated in Figure 7. Similar to DCN in [29], $\mathfrak{R}$ denotes the regular grid receptive field and dilation. For a 3×3 kernel, we have:

$$\mathfrak{R} = \{(-1, -1), (-1, 0), ..., (0, 1), (1, 1)\} \tag{1}$$

The output of AdaptConv is:

$$X_i(p_0) = \sum_{p_n \in \mathfrak{R}} w(p_n) \cdot X_c(p_0 + p_n + \delta p_n) \tag{2}$$

where p_n represents the locations in $\mathfrak{R}$, w denotes the kernel weights, and δp_n is the offset field for each location p_n. In our method, we redefine the offset field δp_n so that DCN can be transformed into a regular convolution with angle information. The offset of AdaptConv is defined as follows:

$$\delta p_i = M_r(\theta) \cdot p_i - p_i \tag{3}$$

where $M_r(\theta) \in \mathbb{R}^{H \times W \times 1}$ is the angle feature information that is split and resized from the location feature information.

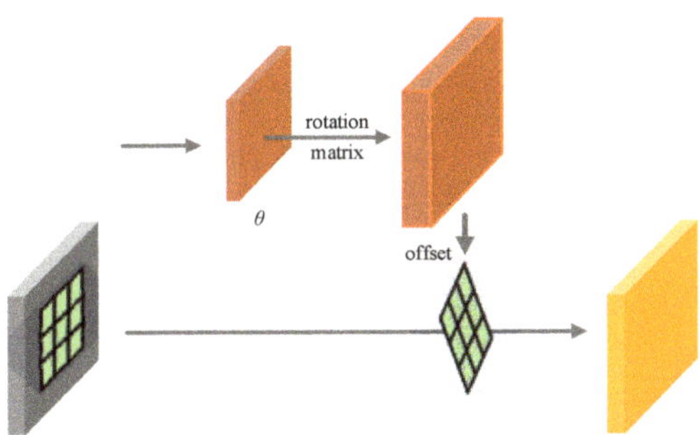

Figure 7. The overall process of AdaptConv. Decoded angle feature map θ is used to generate the offset. The special offset causes the DCN to have a receptive field with regular shape and angle information.

As shown in the bottom of Figure 6, in order to cope with objects with large aspect ratios, we use a three-split AdaptConv with 3×3, 1×3, and 3×1 kernels, which are denoted as $X_i \in \mathbb{R}^{H \times W \times C'}(i \in \{1,2,3\})$, to extract multiple features from $X_c \in \mathbb{R}^{H \times W \times C'}$. In order to cause the receptive fields of neurons to adjust features dynamically, we adopt an attention mechanism to integrate features from the above three-split process. Let the attention map be $A_i \in \mathbb{R}^{H \times W \times 1}(i \in 1,2,3)$ and the computation be as follows:

Firstly, X_i is fed into the attention block, which is composed of Conv1×1 and the batch normalization operation. Secondly, $A_i(i = 1,2,3)$ is sent to SoftMax to obtain the normalized selection weight A'_i:

$$A'_i = SoftMax([A_1, A_2, A_3]) \tag{4}$$

Here, the SoftMax can be described as follows. Suppose v is a vector and v_i represents the i-th element in v. In this case, the SoftMax value of this element is formulated by:

$$p = \frac{e^{v_i}}{\sum_{j=1} e^{v_j}} \tag{5}$$

where the calculation result is between 0 and 1 and the sum of the SoftMax values of all elements is 1.

Thirdly, the feature map Y is obtained by implementing a ReLU operation on:

$$Y = \sum_i A'_i \cdot X_i, \tag{6}$$

where $Y \in \mathbb{R}^{H \times W \times C}$ is the output feature.

The adjusted feature map Y is then sent to the feature refinement module (as shown in the middle of Figure 6) to reconstruct the features and achieve feature alignment. The feature alignment details are illustrated in Figure 8. For each feature map, the aligned feature vectors are obtained through interpolation, according to the five coordinates (orange points) of the refined bounding box. Following the method described in [4], we use feature bilinear interpolation to generate more accurate feature vectors and replace the original feature vectors, as illustrated in Figure 8b. The bilinear interpolation is formulated as follows:

$$val = val_{lt} \times area_{rb} + val_{rt} \times area_{lb}$$
$$+ val_{rb} \times area_{lt} + val_{lb} \times area_{rt}, \tag{7}$$

where val denotes the result of bilinear interpolation. val_{lt}, val_{rt}, val_{rb}, and val_{lb} denote the values of the top-left, top-right, bottom-right, and bottom-left pixel, respectively. $area_{lt}$, $area_{rt}$, $area_{rb}$, and $area_{lb}$ denote the area of the top-left, top-right, bottom-right, and bottom-left rectangles, respectively.

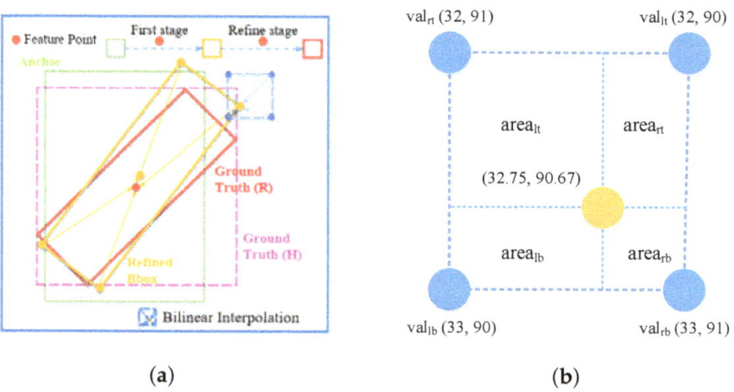

Figure 8. Feature refinement. (**a**) Refine the bounding box with aligned features. (**b**) Feature bilinear interpolation.

3.4. RetinaNet-Based Rotation Detection and Loss Function

We achieve rotated bounding box detection by using the oriented rectangle representation method proposed in [4]. For the completeness of the content, let us introduce the method briefly. We use a vector with five parameters (x, y, w, h, θ) to represent an arbitrarily oriented bounding box, where (x, y) denotes the coordinates of the bounding box center, w and h denote the width and height of the bounding box, and θ denotes the rotation angle of the bounding box relative to the horizontal direction. Compared to the horizontal bounding box, an additional angular offset must be predicted in the regression subnet, for which the rotation bounding box is described as follows:

$$t_x = (x - x_a)/\omega_a, t_y = (y - y_a)/h_a$$
$$t_\omega = \log(\omega/\omega_a), t_h = \log(h/h_a), t_\theta = (\theta - \theta_a) \tag{8}$$

$$t'_x = (x - x_a)/\omega_a, t'_y = (y' - y_a)/h_a$$
$$t'_\omega = \log(\omega'/\omega_a), t'_h = \log(h'/h_a), t'_\theta = (\theta' - \theta_a) \tag{9}$$

where (x, x_a, x') correspond to the ground-truth box, the anchor box, and the predicted box, respectively (likewise for y, w, h, θ).

The definition of the multi-task loss function is as follows:

$$L = \frac{\lambda_1}{N} \sum_{n=1}^{N} t'_n \sum_{j \in \{x,y,w,h,\theta\}} \frac{L_{reg}\left(v'_{nj}, v_{nj}\right)}{\left|L_{reg}\left(v'_{nj}, v_{nj}\right)\right|} \left|-\log(IoU)\right|$$
$$+ \frac{\lambda_2}{h \times w} \sum_{i}^{h} \sum_{j}^{w} L_{att}\left(u'_{nj}, u_{nj}\right) + \frac{\lambda_3}{N} \sum_{n=1}^{N} L_{cls}(p_n, t_n) \tag{10}$$

where N denotes the anchor number and t'_n denotes a binary value ($t'_n = 1$ for the foreground and $t'_n = 0$ for the background). v'_{nj} denotes the predicted offset vectors, and v_{nj} denotes the vector of the ground truth, t_n denotes the instance label, and p_n denotes the probability of the categories calculated by the sigmoid function. The hyperparameters λ_1, λ_2, and λ_3 control the trade-off and are set to 1 by default. The classification loss L_{cls} is implemented using focal loss. In [23], the authors noticed that the imbalance of instances categories results in a low accuracy for a single-stage detector compared with that of a two-stage detector. They proposed focal loss to address this problem. Thus, we use focal loss to optimize our classification loss, whereby our detector maintains single-stage speed while improving the detection accuracy.

Equation (11) shows the cross-entropy loss function that produces focal loss:

$$CE(p_t, y) = -\log(p_t), p_t = \begin{cases} p & if \quad y = 1 \\ 1 - p & otherwise \end{cases} \tag{11}$$

where $y \in \{\pm 1\}$ specifies the ground-truth class and $p_t \in [0, 1]$ is the model's estimated probability for the class with the label $y = 1$.

Furthermore, a weighting factor $\alpha_t \in [0, 1]$ and a modulating factor $(1 - p_t)^\gamma$ ($\gamma \geq 0$) are introduced (as shown in Equation (12)) to control the weights of positive and negative instances, meaning that the training is relatively more focused on positive samples.

$$FL(p_t) = -\alpha_t(1 - p_t)^\gamma \log(p_t) \tag{12}$$

In the rotated object detection task, the loss is very large due to the periodicity of the angle. Therefore, the model has to be regressed in other complex forms, increasing the difficulty of regression. Yang [15] proposed a loss function by introducing the IoU constant factor in the traditional smooth L_1 loss. The smooth L_1 loss is expressed by:

$$Smooth_{L_1}(x) = \begin{cases} 0.5x^2 & |x| < 1 \\ |x| - 0.5 & x < 1 \quad or \quad x > 1 \end{cases} \tag{13}$$

The new regression loss can be divided into two parts, as shown in Equation (10), where $\dfrac{L_{reg}\left(v'_{nj}, v_{nj}\right)}{\left|L_{reg}\left(v'_{nj}, v_{nj}\right)\right|}$ determines the direction of gradient propagation and $|-\log(IoU)|$ determines the magnitude of the gradient.

4. Experiments and Analysis

4.1. Benchmark Datasets

Extensive experiments and ablation studies were conducted. We compared our detector with 8 other well-known detectors through experiments on two challenging satellite optical image benchmarks: DOTA [5] and HRSC2016 [33].

DOTA is the largest and most challenging dataset with both horizontal and oriented bounding box annotations for object detection in satellite or aerial optical images. It contains 2806 satellite images, whose sizes range from 800×800 to 4000×4000. DOTA contains objects with a wide variety of scales, orientations, and appearances. These images have been annotated by experts using 15 common object categories. The object categories include plane (PL), ship (SP), large vehicle (LV), small vehicle (SV), helicopter (HC), tennis court (TC), bridge (BR), ground track field (GTF), basketball court (BC), baseball diamond (BD), soccer field (SBF), storage tank (ST), roundabout (RA), harbor (HA), and swimming pool (SP). Among them, there are huge numbers of densely distributed objects, such as small vehicles, large vehicles, ships, and planes. There are many object categories with large aspect ratios, such as large vehicles, ships, harbors, and bridges. Two detection tasks with horizontal bounding boxes and orientated bounding boxes can be performed on DOTA. In our experiment, we chose the task of detecting objects with an orientated bounding box. An official website (https://captain-whu.github.io/DOTA/dataset.html (accessed on 1

January 2018) is provided for the submission of the results. DOTA contains 1403 training images, 468 verification images, and 935 testing images, which are randomly selected from the original images.

HRSC2016 [33] is a challenging satellite optical imagery dataset for ship detection. It contains 1061 images collected from Google Earth and over 20 categories of ship instances with different shapes, orientations, sizes, and backgrounds. The images with the scenario of ships close to the shore in HRSC2016 were collected from six famous harbors, while the other images show the scenario of ships on the sea. The image size ranges between 300×300 and 1500×900. HRSC2016 contains 436 training images, 181 validation images, and 444 testing images. During the training and testing, we resized the images to 800×800. In our experiment, we chose the task of detecting ships with an orientated bounding box.

4.2. Implementation Details

We adopted ResNet101 FPN as the backbone of the experiment. The hyperparameters of the multi-task loss function were set to $\lambda_1 = 4$, $\lambda_2 = 1$, and $\lambda_3 = 2$. The hyperparameters of the focal loss were set to $\alpha = 0.25$ and $\gamma = 2.0$. SGD [34] was adopted as an optimizer. The initial learning rate was set at 0.04 and the learning rate was divided by 10 at each decay step. The momentum and weight decay were set to 0.9 and 0.0001. The learning rate warmup was set to 500 iterations. We adopted mmdetections [35] as training schedules and trained all the models in 12 epochs for DOTA and 36 epochs for HRSC2016. We used a sever with 4 NVIDIA TITAN Xp GPUs and 4 GPUs with a total batch size of 8 for training and a single GPU for inference.

4.3. Ablation Study

In order to evaluate the impact of DFR, FPT, and data augmentation on our detector, we conducted some ablation studies on the DOTA and HRSC2016. ResNet-50 pretrained on ImageNet was used as a backbone in the experiments. The weight decay and momentum were set to 0.0001 and 0.9, respectively. Detectors were trained using 4 GPUs with a total of 8 images per mini batch (two images per GPU).

4.3.1. Ablation Study for DFR

In this subsection, we present the ablation study results for the original feature refinement module (FRM) and the proposed DFR. As shown in Table 1, RetinaNet has a 62.22% accuracy. By introducing FRM, R^{3}Det (RetinaNet with refinement) obtained a 71.69% accuracy under ResNet101-FPN as a backbone with no multi-scale. FRM improved the accuracy by 9.47%. In this study, we introduced DFR to achieve feature misalignment instead of FRM. The accuracy with DFR was 73.10%, which is 1.41% higher then the accuracy with FRM. As shown in Table 2, the accuracy for some hard instance categories, such as BR, SV, LV, SH, and RA, increased by 2.06%, 7.71%, 2.8%, 9.42%, and 2.84%, respectively. We can see that the proposed DFR has a significant effect on improving the performance.

Table 1. Ablation study of DFR, FPT, and data augmentation.

Methods	mAP	FRM	DFR	FPT	Data Aug.
RetinaNet [23]	62.22	×	-	-	-
R^{3}Det [4]	71.69	√	-	-	-
	73.10	-	√	×	×
ADT-Det (ours)	73.77	-	√	√	×
	76.89	-	√	√	√

Table 2. Ablation study of FRM and the proposed DFR, where FRM is the original feature refinement module proposed by R^{3}Det.

Methods	PL	BD	BR	GTF	SV	LV	SH	TC	BC	ST	SBF	RA	HA	SP	HC	mAP
FRM	89.54	81.99	48.46	62.52	70.48	74.29	77.54	90.80	81.39	83.54	61.79	59.82	65.44	67.46	60.05	71.69
DFR	88.99	79.42	50.52	68.62	78.19	77.09	86.96	90.85	79.82	85.45	58.99	62.66	66.01	67.56	55.45	73.10

4.3.2. Ablation Study on FPT

As shown in Table 1, the accuracy was 73.10% without FPT and 73.77% with FPT. It can be seen that the proposed FPT has a slight effect on improving the performance.

4.3.3. Ablation Study for Data Augmentation

A previous study showed that data augmentation is a very effective way to improve detection performance by enriching training datasets. In this subsection, we study the impact of data augmentation on the detection accuracy of our detector. The data augmentation methods used in the experiment includes horizontal and vertical flipping, random graying, multi-scales, and random rotation. As shown in Table 1, the detection accuracy was improved from 73.77% to 76.89% by data augmentation.

4.4. Comparison to State of the Art

4.4.1. Results on DOTA

We compared our proposed detector with some state-of-the-art detectors using the DOTA dataset. The results reported here were obtained by submitting our detection results to the official DOTA evaluation server. All the detectors involved in this experiment can be divided into three groups: multi-stage, anchor-free, and single-stage detectors. As shown in Table 3, the latest multi-stage detectors, such as SCRDet [15], Gliding Vertex [10], and APE [36], achieved values of 69.56%, 72.61%, 75.02%, and 75.75% mAP, respectively. The anchor-free method DRN [32] achieved a 73.23% mAP. The single-stage detectors R^3Det and R^4Det with ResNet-152 had 73.73% and 75.84% accuracies. Our ADT-Det with ResNet-152 achieved the highest accuracy of 77.43%, which is 1.59% higher than the previous best result.

The research of R^4Det [3] showed that feature recursion is a good method to improve the detection accuracy. We also adopted feature recursion in our pipeline, and it outperformed state-of-art methods and achieved a 79.95% accuracy.

The visualization of some of the detection results of our detector is shown in Figure 9. The results demonstrate that our detector can accurately detect most objects with arbitrary orientations, large aspect ratios, huge scale differences, and dense distributions.

Figure 9. Visualization of some detection results on DOTA. Different colored bounding boxes represent instances of different categories (best viewed in color).

Table 3. Detection accuracy on different objects (AP) and overall performance (mAP) evaluation on DOTA.

Methods	PL	BD	BR	GTF	SV	LV	SH	TC	BC	ST	SBF	RA	HA	SP	HC	mAP
Two-stage methods																
R-FCN [12]	37.80	38.21	3.64	37.26	6.74	2.60	5.59	22.85	46.93	66.04	33.37	47.15	10.60	25.19	17.96	26.79
FR-H [5]	47.16	61.00	9.80	51.74	14.87	12.80	6.88	56.26	59.97	57.32	47.83	48.70	8.23	37.25	23.05	32.29
FR-O [5]	79.09	69.12	17.17	63.49	34.20	37.16	36.20	89.19	69.60	58.96	49.40	52.52	46.69	44.80	46.30	52.93
IE-Net [37]	80.20	64.54	39.82	32.07	49.71	65.01	52.58	81.45	44.66	78.51	46.54	56.73	64.40	64.24	36.75	57.14
R^{2}CNN [11]	80.94	65.67	35.34	67.44	59.92	50.91	55.81	90.67	66.92	72.39	55.06	52.23	55.14	53.35	48.22	60.67
RoI-Transformer [2]	88.64	78.54	43.44	75.92	68.81	73.68	83.59	90.74	77.27	81.46	58.39	53.54	62.83	58.93	47.67	69.56
SCRDet [15]	89.98	80.65	52.09	68.36	68.83	60.36	72.41	90.85	**87.94**	86.86	65.02	66.68	66.25	68.24	65.21	72.61
RSDet [4]	**90.10**	82.00	53.80	68.5	70.20	78.7	73.6	91.2	87.1	84.7	64.31	68.2	66.1	69.3	63.7	74.1
Gliding Vertex [10]	89.64	**85.00**	52.26	77.34	73.01	73.14	86.82	90.74	79.02	86.81	59.55	70.91	72.94	70.86	57.32	75.02
FFA [38]	90.10	82.70	54.20	75.20	71.00	79.90	83.50	90.70	83.90	84.60	61.20	68.0	70.70	76.00	63.70	75.00
APE [36]	89.96	83.64	53.42	76.03	74.01	77.16	79.45	90.83	87.15	84.51	67.72	60.33	74.61	71.84	65.55	75.75
Anchor-free methods																
DRN [32]	89.71	82.34	47.22	64.10	76.22	74.43	85.84	90.57	86.18	84.89	57.65	61.93	69.30	69.63	58.48	73.23
Single-stage methods																
SSD [21]	39.57	9.09	0.64	13.18	0.26	0.39	1.11	16.24	27.57	9.23	27.16	9.09	3.03	1.05	1.01	10.59
YOLO v2 [19]	39.49	20.29	36.58	23.42	8.85	2.09	4.82	44.34	38.35	34.65	16.02	37.62	47.23	25.5	7.45	21.39
R^{3}Det [4]-ResNet152	89.49	81.17	5.53	66.10	70.92	78.66	78.21	90.81	85.26	84.23	61.81	63.77	68.16	68.83	67.17	73.73
R^{4}Det [3]-ResNet152	88.96	85.42	52.91	73.84	74.86	81.52	80.29	90.79	86.95	85.25	64.05	60.93	69.00	70.55	**67.76**	75.84
ADT-Det (no Multi-Scale Training)	88.99	79.42	50.52	68.62	78.19	77.09	86.96	90.85	79.82	85.45	58.99	62.66	66.01	67.56	55.45	73.10
ADT-Det-ResNet50	89.28	83.97	51.44	79.12	78.31	82.18	87.79	90.82	84.84	87.46	65.47	64.23	71.87	71.40	65.08	76.89
ADT-Det-ResNet101	89.62	84.70	51.88	77.43	77.88	80.54	88.22	90.85	84.18	86.68	66.30	69.17	76.34	70.91	63.01	77.18
ADT-Det-ResNet152	89.61	84.59	53.18	**81.05**	78.31	80.86	88.22	90.82	84.80	**86.89**	69.97	66.78	76.18	72.10	60.03	77.43
ADT-Det (with Feature Recursion)	89.71	84.71	**59.63**	80.94	**80.30**	**83.53**	**88.94**	**90.86**	87.06	87.81	**70.72**	**70.92**	**78.66**	**79.40**	65.99	**79.95**

4.4.2. Result on HRSC2016

HRSC2016 contains many ship instances with large aspect ratios and arbitrary orientations. RRPN was originally developed for orientation scene text detection. RoI-Transformer and R^{3}Det are advanced satellite optical imagery detection methods. We performed comparative experiments with these methods, and the results are shown in Table 4. We can see that the scene text detection methods have competitive results for satellite optical imagery datasets; RRPN [13] achieved a 79.08% mAP. Under the PASCAL VOC2007 metrics, the famous multi-stage rotated object detector RoI-Transformer [2] could achieve an 86.20% accuracy. The state-of-art single-stage methods, R^{3}Det [4] and R^{4}Det [3], could achieve 89.26% and 89.56% accuracies, respectively. Meanwhile, the proposed ADT-Det detector achieved the best detection performance, with an accuracy of 89.75%. This accuracy is close to the accuracy for ship detection in the DOTA experiment (88.94%), which further proves the advantage of using DFR to reduce the mismatch between arbitrarily oriented objects and axis-aligned receptive fields. Evaluated under the PASCAL VOC2012 metrics, the anchor-free method DRN achieved a 92.7% accuracy, while the proposed ADT-Det detector (with ResNet-152) achieved the best detection result, with an accuracy of 93.47%.

4.4.3. Speed Comparison

Comparison experiments for detection speed and accuracy were carried out on HRSC2016. In the experiment, our ADT-Det detector was compared with eight other well-known methods. The detailed results are illustrated in Table 4 and the overall comparison results are also visualized in Figure 10. It can be seen that the multi-stage detector RoI-Transformer could achieve an 86.2% accuracy and a 6 fps speed when using ResNet101 as the backbone and when the input image size was 512 × 800. The single-stage R^{3}Det detector could achieve a 89.26% accuracy and a 10 fps speed. The existing state-of-art single-stage R^{4}Det could achieve an 89.5% accuracy, but the detection speed was slower than that of R^{3}Det. Our ADT-Det detector could achieve an 89.75% accuracy when evaluated under the PASCAL VOC2007 metrics and a 12 fps speed when the input image size was 800 × 800. Furthermore, we could achieve a 14.6 fps speed when the input image size was 600 × 600. The results demonstrate that our ADT-Det detector can achieve the highest accuracy of all the investigated detectors while running very fast.

Table 4. Evaluation results with the accuracy and speed of some well-known detectors on HRSC2016. All models were evaluated under ResNet-152. * indicates that the result was evaluated under the PASCAL VOC2012 metrics.

Methods	RC1&RC2 [39]	RRPN [13]	RRD [40]	RoI-Trans. [2]	DRN [32]	CenterMap-Net [41]	R^{3}Det [4]	R^{4}Det [3]	ADT-Det	
Input size	300 × 300	800 × 800	384 × 384	512 × 800	768 × 768	768 × 768	800 × 800	800 × 800	600 × 600	80 0× 800
AP	75.7	79.08	84.3	86.20	92.7 *	92.8 *	89.26	89.56	88.96	89.75/93.47 *
Speed	Slow(<1 fps)	3.5fps	Slow(<1 fps)	6 fps	-	-	10 fps	6.5 fps	14.6 fps	12 fps

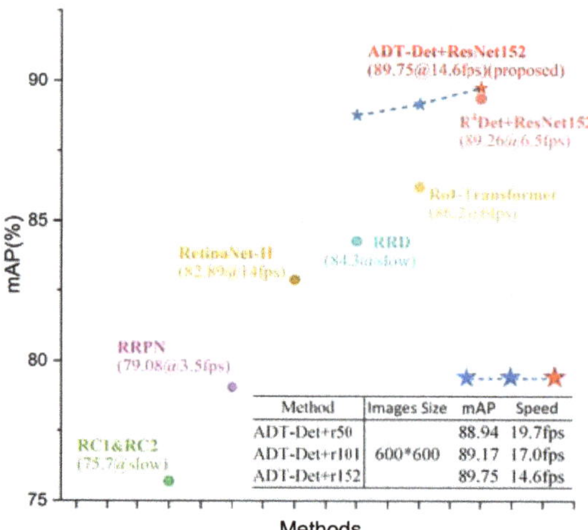

Method	Images Size	mAP	Speed
ADT-Det+r50		88.94	19.7fps
ADT-Det+r101	600*600	89.17	17.0fps
ADT-Det+r152		89.75	14.6fps

Figure 10. Detection performance (mAP) and speed comparison of our ADT-Det detector and 5 other famous detectors on HRSC2016. Our ADT-Det detector achieved the highest accuracy of all the investigated detectors while running very fast. Detailed results are listed in Table 4.

5. Conclusions

In this work, we identify inappropriate feature extraction as the primary obstacle preventing the high-performance detection of instances with arbitrary directions, large aspect ratios, and dense distributions. To address this, we proposed the use of an adaptive dynamic refined single-stage transformer detector to address the aforementioned challenges, aiming to achieve a high recall and speed. Our detector realizes rotated object detection with RetinaNet as the baseline to achieve the detection of multi-scale objects and densely distributed objects. Firstly, the feature pyramid transformer (FPT) was introduced into the traditional feature pyramid network (FPN) to enhance feature extraction through a feature interaction mechanism. Secondly, the output features of FPT were fed into two post-processing steps, considering the mismatch between the rotated bounding box and the general axis-aligned receptive fields of CNN. Dynamic Feature Refinement (DFR) was introduced in the refinement step. The key idea of DFR was to adaptively adjust the feature map and reconstruct a new feature map for arbitrary-oriented object detection to alleviate the mismatches between the rotated bounding box and the axis-aligned receptive fields. Extensive experiments and ablation studies were carried out to test the proposed detector based on two challenging satellite optical imagery public datasets, DOTA and HRSC2016. The proposed detector could achieve a 79.95% mAP accuracy for DOTA and 93.47% mAP for HRSC2016, and the running speed was 14.6 fps with an 600×600 input image size. The results show that our method achieved state-of-the-art results in the task of object detection in these optical imagery datasets.

Author Contributions: The first two authors have equally contributed to the work. Conceptualization, Y.Z.; methodology, Y.Z., P.S. and Z.Z.; software, P.S.; validation, W.X., Q.Z.; formal analysis, Y.Z., P.S. and Z.Z.; investigation, Y.Z., P.S. and W.X.; resources, Y.Z. and Z.Z.; writing—original draft preparation, Y.Z. and P.S.; writing—review and editing, Z.Z., W.X. and Q.R.; visualization, P.S. and Q.R.; supervision, Y.Z. and Z.Z.; project administration, Y.Z. and Z.Z.; funding acquisition, Z.Z. All authors have read and agreed to the published version of the manuscript.

Funding: This work was supported in part by the National Natural Science Foundation of China under Grant 61403412.

Institutional Review Board Statement: Not applicable.

Informed Consent Statement: Not applicable.

Data Availability Statement: The data presented in this study are available on request from the corresponding author. The data are not publicly available due to their large size.

Conflicts of Interest: The authors declare no conflict of interest.

References

1. Liu, L.; Ouyang, W.; Wang, X.; Fieguth, P.; Chen, J.; Liu, X.; Pietikäinen, M. Deep learning for generic object detection: A survey. *Int. J. Comput. Vis.* **2020**, *128*, 261–318. [CrossRef]
2. Ding, J.; Xue, N.; Long, Y.; Xia, G.S.; Lu, Q. Learning roi transformer for oriented object detection in aerial images. In Proceedings of the IEEE/CVF Conference on Computer Vision and Pattern Recognition, Long Beach, CA, USA, 16–20 June 2019; pp. 2849–2858.
3. Sun, P.; Zheng, Y.; Zhou, Z.; Xu, W.; Ren, Q. R4 Det: Refined single-stage detector with feature recursion and refinement for rotating object detection in aerial images. *Image Vis. Comput.* **2020**, *103*, 104036. [CrossRef]
4. Yang, X.; Liu, Q.; Yan, J.; Li, A.; Zhang, Z.; Yu, G. R3det: Refined single-stage detector with feature refinement for rotating object. *arXiv* **2019**, arXiv:1908.05612.
5. Xia, G.S.; Bai, X.; Ding, J.; Zhu, Z.; Belongie, S.; Luo, J.; Datcu, M.; Pelillo, M.; Zhang, L. DOTA: A large-scale dataset for object detection in aerial images. In Proceedings of the IEEE Conference on Computer Vision and Pattern Recognition, Salt Lake City, UT, USA, 18–22 June 2018; pp. 3974–3983.
6. Cheng, G.; Zhou, P.; Han, J. Learning rotation-invariant convolutional neural networks for object detection in VHR optical remote sensing images. *IEEE Trans. Geosci. Remote Sens.* **2016**, *54*, 7405–7415. [CrossRef]
7. Liu, Z.; Wang, H.; Weng, L.; Yang, Y. Ship rotated bounding box space for ship extraction from high-resolution optical satellite images with complex backgrounds. *IEEE Geosci. Remote Sens. Lett.* **2016**, *13*, 1074–1078. [CrossRef]
8. Zhang, G.; Lu, S.; Zhang, W. CAD-Net: A context-aware detection network for objects in remote sensing imagery. *IEEE Trans. Geosci. Remote Sens.* **2019**, *57*, 10015–10024. [CrossRef]
9. Hou, J.B.; Zhu, X.; Yin, X.C. Self-Adaptive Aspect Ratio Anchor for Oriented Object Detection in Remote Sensing Images. *Remote Sens.* **2021**, *13*, 1318. [CrossRef]
10. Xu, Y.; Fu, M.; Wang, Q.; Wang, Y.; Chen, K.; Xia, G.S.; Bai, X. Gliding vertex on the horizontal bounding box for multi-oriented object detection. *IEEE Trans. Pattern Anal. Mach. Intell.* **2020**, *43*, 1452–1459. [CrossRef] [PubMed]
11. Jiang, Y.; Zhu, X.; Wang, X.; Yang, S.; Li, W.; Wang, H.; Fu, P.; Luo, Z. R2CNN: Rotational Region CNN for Orientation Robust Scene Text Detection. *arXiv* **2017**, arXiv:1706.09579.
12. Dai, J.; Li, Y.; He, K.; Sun, J. R-fcn: Object detection via region-based fully convolutional networks. *arXiv* **2016**, arXiv:1605.06409.
13. Ma, J.; Shao, W.; Ye, H.; Wang, L.; Wang, H.; Zheng, Y.; Xue, X. Arbitrary-oriented scene text detection via rotation proposals. *IEEE Trans. Multimed.* **2018**, *20*, 3111–3122. [CrossRef]
14. Li, Y.; Huang, Q.; Pei, X.; Jiao, L.; Shang, R. RADet: Refine Feature Pyramid Network and Multi-Layer Attention Network for Arbitrary-Oriented Object Detection of Remote Sensing Images. *Remote Sens.* **2020**, *12*, 389. [CrossRef]
15. Yang, X.; Yang, J.; Yan, J.; Zhang, Y.; Zhang, T.; Guo, Z.; Sun, X.; Fu, K. Scrdet: Towards more robust detection for small, cluttered and rotated objects. In Proceedings of the IEEE/CVF International Conference on Computer Vision, Long Beach, CA, USA, 16–20 June 2019; pp. 8232–8241.
16. Lee, J.; Kim, D.; Ponce, J.; Ham, B. Sfnet: Learning object-aware semantic correspondence. In Proceedings of the IEEE/CVF Conference on Computer Vision and Pattern Recognition, Long Beach, CA, USA, 16–20 June 2019; pp. 2278–2287.
17. Ren, S.; He, K.; Girshick, R.; Sun, J. Faster R-CNN: Towards real-time object detection with region proposal networks. *IEEE Trans. Pattern Anal. Mach. Intell.* **2016**, *39*, 1137–1149. [CrossRef] [PubMed]
18. Redmon, J.; Divvala, S.; Girshick, R.; Farhadi, A. You only look once: Unified, real-time object detection. In Proceedings of the IEEE Conference on Computer Vision and Pattern Recognition, Las Vegas, NV, USA, 26–30 June 2016; pp. 779–788.
19. Redmon, J.; Farhadi, A. YOLO9000: Better, faster, stronger. In Proceedings of the IEEE Conference on Computer Vision and Pattern Recognition, Venice, Italy, 22–29 October 2017; pp. 7263–7271.
20. Redmon, J.; Farhadi, A. Yolov3: An incremental improvement. *arXiv* **2018**, arXiv:1804.02767.
21. Liu, W.; Anguelov, D.; Erhan, D.; Szegedy, C.; Reed, S.; Fu, C.Y.; Berg, A.C. Ssd: Single shot multibox detector. In *European Conference on Computer Vision*; Springer: Berlin/Heidelberg, Germany, 2016; pp. 21–37.
22. Neubeck, A.; Van Gool, L. Efficient non-maximum suppression. In Proceedings of the 18th International Conference on Pattern Recognition (ICPR'06), Hong Kong, China, 20–24 August 2006; Volume 3, pp. 850–855.
23. Lin, T.Y.; Goyal, P.; Girshick, R.; He, K.; Dollár, P. Focal loss for dense object detection. In Proceedings of the IEEE International Conference on Computer Vision, Venice, Italy, 22–29 October 2017; pp. 2980–2988.
24. Lin, T.Y.; Dollár, P.; Girshick, R.; He, K.; Hariharan, B.; Belongie, S. Feature pyramid networks for object detection. In Proceedings of the IEEE Conference on Computer Vision and Pattern Recognition, Venice, Italy, 22–29 October 2017; pp. 2117–2125.
25. Zhang, D.; Zhang, H.; Tang, J.; Wang, M.; Hua, X.; Sun, Q. Feature pyramid transformer. In *European Conference on Computer Vision*; Springer: Berlin/Heidelberg, Germany, 2020; pp. 323–339.
26. Yu, F.; Koltun, V. Multi-scale context aggregation by dilated convolutions. *arXiv* **2015**, arXiv:1511.07122.
27. Chen, L.C.; Zhu, Y.; Papandreou, G.; Schroff, F.; Adam, H. Encoder-decoder with atrous separable convolution for semantic image segmentation. In Proceedings of the European Conference on Computer Vision (ECCV), Munich, Germany, 8–14 September, 2018; pp. 801–818.

28. Yang, M.; Yu, K.; Zhang, C.; Li, Z.; Yang, K. Denseaspp for semantic segmentation in street scenes. In Proceedings of the IEEE Conference on Computer Vision and Pattern Recognition, Salt Lake City, UT, USA, 18–22 June 2018; pp. 3684–3692.

29. Dai, J.; Qi, H.; Xiong, Y.; Li, Y.; Zhang, G.; Hu, H.; Wei, Y. Deformable convolutional networks. In Proceedings of the IEEE International Conference on Computer Vision, Venice, Italy, 22–29 October 2017; pp. 764–773.

30. Cai, Z.; Vasconcelos, N. Cascade r-cnn: Delving into high quality object detection. In Proceedings of the IEEE Conference on Computer Vision and Pattern Recognition, Salt Lake City, UT, USA, 18–22 June 2018; pp. 6154–6162.

31. Zhang, S.; Wen, L.; Bian, X.; Lei, Z.; Li, S.Z. Single-shot refinement neural network for object detection. In Proceedings of the IEEE Conference on Computer Vision and Pattern Recognition, Salt Lake City, UT, USA, 18–22 June 2018; pp. 4203–4212.

32. Pan, X.; Ren, Y.; Sheng, K.; Dong, W.; Yuan, H.; Guo, X.; Ma, C.; Xu, C. Dynamic refinement network for oriented and densely packed object detection. In Proceedings of the IEEE/CVF Conference on Computer Vision and Pattern Recognition, Virtual Conference, 16–18 June 2020; pp. 11207–11216.

33. Liu, Z.; Yuan, L.; Weng, L.; Yang, Y. A high resolution optical satellite image dataset for ship recognition and some new baselines. In Proceedings of the International Conference on Pattern Recognition Applications and Methods, SCITEPRESS, Porto, Portugal, 24–26 February 2017; Volume 2, pp. 324–331.

34. Bottou, L. Stochastic gradient descent tricks. In *Neural Networks: Tricks of the Trade*; Springer: Berlin/Heidelberg, Germany, 2012; pp. 421–436.

35. Chen, K.; Wang, J.; Pang, J.; Cao, Y.; Xiong, Y.; Li, X.; Sun, S.; Feng, W.; Liu, Z.; Xu, J.; et al. MMDetection: Open MMLab Detection Toolbox and Benchmark. *arXiv* **2019**, arXiv:1906.07155.

36. Zhu, Y.; Du, J.; Wu, X. Adaptive period embedding for representing oriented objects in aerial images. *IEEE Trans. Geosci. Remote Sens.* **2020**, *58*, 7247–7257. [CrossRef]

37. Lin, Y.; Feng, P.; Guan, J. IENet: Interacting embranchment one stage anchor free detector for orientation aerial object detection. *arXiv* **2019**, arXiv:1912.00969.

38. Qin, X.; Wang, Z.; Bai, Y.; Xie, X.; Jia, H. FFA-Net: Feature fusion attention network for single image dehazing. In Proceedings of the AAAI Conference on Artificial Intelligence, Hilton, New York, NY, USA, 7–12 February 2020; Volume 34, pp. 11908–11915.

39. Liu, L.; Pan, Z.; Lei, B. Learning a rotation invariant detector with rotatable bounding box. *arXiv* **2017**, arXiv:1711.09405.

40. Liao, M.; Zhu, Z.; Shi, B.; Xia, G.s.; Bai, X. Rotation-sensitive regression for oriented scene text detection. In Proceedings of the IEEE Conference on Computer Vision and Pattern Recognition, Salt Lake City, UT, USA, 18–22 June 2018; pp. 5909–5918.

41. Wang, J.; Yang, W.; Li, H.C.; Zhang, H.; Xia, G.S. Learning Center Probability Map for Detecting Objects in Aerial Images. *IEEE Trans. Geosci. Remote Sens.* **2020**, *5*, 4307–4323. [CrossRef]

remote sensing

Article

A Multiview Semantic Vegetation Index for Robust Estimation of Urban Vegetation Cover

Asim Khan [1,*], Warda Asim [1], Anwaar Ulhaq [1,2] and Randall W. Robinson [1,3]

[1] The Institute for Sustainable Industries and Liveable Cities (ISILC), College of Engineering and Science, Victoria University, Melbourne, VIC 8001, Australia; asim.khan@vu.edu.au (A.K); warda.asim@live.vu.edu.au (W.A.); aulhaq@csu.edu.au (A.U.); randall.robinson@vu.edu.au (R.W.R.)
[2] School of Computing, Mathematics and Engineering, Charles Sturt University, Port Macquarie, NSW 2444, Australia
[3] Applied Ecology Research Group, The Institute for Sustainable Industries and Liveable Cities (ISILC), College of Engineering and Science, Victoria University, Melbourne, VIC 8001, Australia
* Correspondence: asim.khan@vu.edu.au

Abstract: Urban vegetation growth is vital for developing sustainable and liveable cities in the contemporary era since it directly helps people's health and well-being. Estimating vegetation cover and biomass is commonly done by calculating various vegetation indices for automated urban vegetation management and monitoring. However, most of these indices fail to capture robust estimation of vegetation cover due to their inherent focus on colour attributes with limited viewpoint and ignore seasonal changes. To solve this limitation, this article proposed a novel vegetation index called the Multiview Semantic Vegetation Index (MSVI), which is robust to color, viewpoint, and seasonal variations. Moreover, it can be applied directly to RGB images. This Multiview Semantic Vegetation Index (MSVI) is based on deep semantic segmentation and multiview field coverage and can be integrated into any vegetation management platform. This index has been tested on Google Street View (GSV) imagery of Wyndham City Council, Melbourne, Australia. The experiments and training achieved an overall pixel accuracy of 89.4% and 92.4% for FCN and U-Net, respectively. Thus, the MSVI can be a helpful instrument for analysing urban forestry and vegetation biomass since it provides an accurate and reliable objective method for assessing the plant cover at street level.

Keywords: multiview semantic vegetation index; urban forestry; green view index (GVI); semantic segmentation; urban vegetation; RGB vegetation index

Citation: Khan, A.; Asim, W.; Ulhaq, A.; Robinson, R.W A Multiview Semantic Vegetation Index for Robust Estimation of Urban Vegetation Cover. *Remote Sens.* **2022**, *14*, 228. https://doi.org/10.3390/rs14010228

Academic Editor: Tania Stathaki

Received: 16 October 2021
Accepted: 20 December 2021
Published: 5 January 2022

Publisher's Note: MDPI stays neutral with regard to jurisdictional claims in published maps and institutional affiliations.

1. Introduction

The changing land use patterns and population growth have had a significant impact on the vegetation composition in the world [1–3] which is essential for better living conditions of city dwellers. As indicated by Wolf, K.L. [4], a city's vegetation cover (i.e. street woods, lawns, etc.) has long been acknowledged as a key component of urban landscape planning. According to Appleyard [5], the instrumental role of street vegetation is to absorb airborne pollutants through carbon sequestration and oxygen production, to mitigate noise pollution in urban heat islands [6], and to reduce storm waters [7,8]. In addition, the life of vegetation generally raises the aesthetic evaluation of people in urban settings [9,10]. For this purpose, it is critical to document changes in vegetation so that land management professionals may work to improve the urban environment. Furthermore, changes in the type of land cover (such as building developments) have been found to have a strong correlation with the changes of vegetation in the urban environment.

Moreover, changes in an urban environment are generally very important. Food, energy, water, and land used by urban residents have a significant impact on the environment. Therefore, automated detection of vegetation cover is often done through calculations of various vegetation indexes [11] that hold important information regarding vegetation cover of a particular location. In the past, various algorithms were employed for the calculation of the vegetation index using various image modalities. However, existing approaches have highly focused on spectral analysis and color variations. For instance, Normalized Difference Vegetation Index (NDVI) tends to amplify atmospheric noise in the Near Infrared Reflectance (NIR) and Red bands and becomes very sensitive to background variation. Therefore, it does not work well for RGB images for street-level vegetation analysis. Remote sensing data collected from above by sensors (aircraft, space) misses the glimpse of urban flora at street level. Thus, profile views of urban greenery from the road level are insufficiently assessed, even though green indices derived from remotely sensed image data might help quantify urban greenery. There is a distinction between vegetation view through ground experience and the view captured by remote sensing systems [12]. Li et al. [13,14] discovered that people had unequal access to distinct types of urban greenery (street vegetation, private yard total vegetation, private yard trees and shrubs, and urban parks), providing the groundwork for subsequent research into urban greenery inequity.

On the other hand, RGB based vegetation indexes are prone to wrong estimations due to reliance on green color and ineffectiveness to capture seasonal variations. Rencai et al. [15] utilise the green view index (GVI) as a quantitative indicator to determine how much greenery can be seen by pedestrians and then apply an image segmentation algorithm to figure out how much greenery can be seen by pedestrians in street view images. Zhang et al. [16] used an extensive street view image data set, as well as a horizontal green view index (HGVI), to calculate the quantity of greenery visible from the street in their research. Long et al. [17] analysed 245 Chinese cities, calculating the GVI values of their central regions and comparing them to the overall GVI conditions of the respective cities. As a result, they discovered that more affluent and well-run cities have longer and greener streets. Several visual qualities of streets such as salient region saturation, visual entropy, a green view index, and a sky-openness index were measured by Cheng et al. [18].

Kendal et al. [19] used color threshold for extraction of the vegetation index. The technique proved to be promising, but only using color features for segmentation is not an efficient model as any clutter information in the image can match the vegetation color. Further, in recent years, Bawden et al. and Kattenborn et al [20,21] used convolutional neural network (CNN) for two studies: In the first approach, they used a CNN-based approach to train data acquired from unmanned aerial vehicle (UAV)-based high-resolution RGB imagery visual interpretation, a fine-grained map for two species of vegetation. In the second approach, they mapped species of trees or plants cover in different vegetation UAV RGB imagery. However, these approaches suffer due to reliance on color and specific image features and are unable to handle large variations in vegetation characteristics.

Recent advancements in deep learning have introduced a new level of accuracy in identifying objects of interest through semantic segmentation. Jonathan et al. [22] introduced a fully convolutional neural network (FCN), and Dvornik et al. [23] proposed BlitzNet for object segmentation. Yi et al. [24] constructed an instance aware based semantic segmentation model, which utilized the advantages of FCN for segmentation and classification. As a result of the development, the model was capable of simultaneously recognizing and segmenting the object instances. Liang-Chieh et al. [25] applied fully convolutional neural networks (FCN) to a multi-scale input image in order to achieve the required results.

Motivated with the success of deep semantic segmentation, the conducted research proposes a semantic vegetation index (SVI) for RGB images with robustness against color changes and seasonal variations. To deal with the limitation of single image coverage, its extension, called multiview semantic vegetation index (MSVI), is also introduced, which can estimate vegetation cover from multiple views. The overall framework of this study is presented graphically in Figure 1.

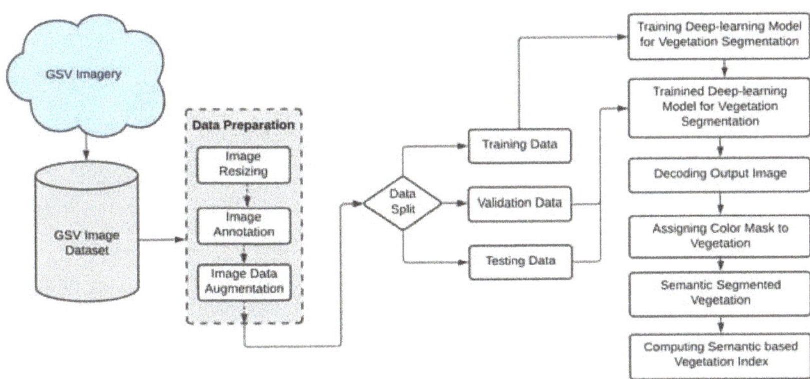

Figure 1. A data flow diagram for the MSVI, which highlights the process of calculating the proposed vegetation index.

The contribution: According to the literature, the semantic vegetation index (SVI) is one of the first approaches to integrate deep semantic segmentation into the process of vegetation index estimation. Although there are a variety of vegetation indexes in the literature, they are limited to a specific image modality and color feature, or they overlook essential flora semantic information. It makes them more susceptible to noise, resulting in erroneous estimation. The proposed index is robust to color and seasonal variations and works for any imaging modality. Furthermore, it can be extended to multiple views to expand exposure and reliable calculation. The segmentation approach is not claimed to have made a contribution in this study. Nonetheless, it compares many ways to determine which are the most appropriate for this aim.

The rest of the paper is organized as follows: Section 2 explains the materials and methods taken into account, Section 3 presents detailed information regarding the experiments and results achieved by the proposed methodology, Section 4 presents the comparative analysis with the previous work , Section 5 presents a detailed discussion of the proposed work, while Section 6 is the conclusion section of this paper.

2. Materials and Methods

2.1. Study Area

Figure 2 shows the municipal council of Wyndham (VIC, Australia), as the selected area for this study. It lies on the western outskirts of Melbourne (VIC, Australia) and covers an area of 542 km^2. According to the 2019 census, its estimated population is 270,478. Wyndham is the third fastest-growing council in the state of Victoria. The population of Wyndham is diverse, and the community development projects suggest that by 2031 more than 330,000 people are expected to come and live. Wyndham is home to 16 suburbs (Cocoroc, Eynesbury, Hoppers Crossing, Laverton North, Laverton RAAF, Little River, Mambourin, Mount Cottrell, Point Cook, Quandong, Tarneit, Truganina, Williams Landing, Manor Lakes, Quandong and Werribee South). The City Council of Wyndham is committed to improving residents' environment and livelihoods. Every year, thousands of new trees and vegetation are planted in response to this commitment to increase Wyndham's tree canopy cover through the street tree planting program [26].

Figure 2. The research area in Victoria, Australia, which was chosen for this study. (**a**) Victoria (Australia); (**b**) Wyndham City Council, Victoria, Australia; (**c**) one sample site and (**d**) a sample street view from a sample site.

2.2. Input Data Set/Google Street View Image Collection

In this research work, Google street view images (GSV) [27] is used for the multiview semantic vegetation index (MSVI) estimation. A sample GSV image of a Wyndham Council in Melbourne, Victoria, is shown in Figure 3. The GSV panorama view is identical to the real-world view. The process of producing a 360° GSV panorama is to sequentially capture horizontal X-number (X = 6) images and vertical Y-numbers (Y = 3) images of the camera [28]. The GSV Image API (Google) [27], together with the position and travelling direction of the GSV car, can be used to obtain every accessible GSV image in an HTTP URL form, for example (https://maps.googleapis.com/maps/api/streetview?parameters (accessed on 15 August 2021). The static GSV image, as shown in Figure 4, can be retrieved for every point where the GSV is available by establishing URL parameters supplied via a specific HTTP request utilising the GSV Image API (Google) [27].

Figure 3. A sample panorama image of a selected study site from Google street view imagery.

The GSV images for each sample site in six directions were collected as illustrated in Figure 5a, and in three vertical angles to determine the green areas visible to pedestrians as presented in Figure 5b. Therefore, 0°, 60°, 120°, 180°, 240°, and 300° were set as the heading parameters whereas 45°, 0°, and −45° as pitch parameters. As a result, a total of eighteen images are captured for a specific location, ensuring that no vegetation area is left out of the index calculation. A Python programming language script is executed on all the GSV images to read and download them from each example site by automatically parsing the GSV URL.

Figure 4. A static image of a research site taken from Google Street View imagery.

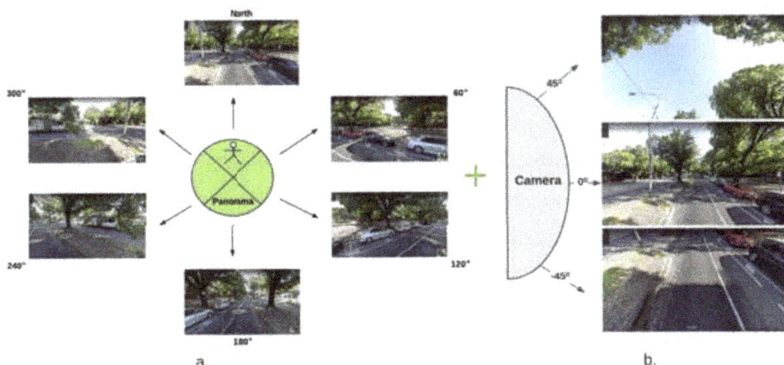

Figure 5. (**a**) Sample of images taken from pedestrian view in six different angles and (**b**) from pedestrian view, three images taken from three vertical angles (45°, 0°, −45°).

2.3. Deep Semantic Segmentation

The act of grouping sections of an image in such a way that each pixel in a group correlates to the object class represented by the group as a whole can be defined as semantic segmentation for images in this manner [29,30]. The object classes in the current work correspond to trees and green vegetation terrain. Images can be segmented by allocating each pixel of an input image to a label class object, which is referred to as semantic image labelling [31]. Image segmentation is also known as semantic image labelling. This method often combines image segmentation with object identification techniques to produce a final result. Various deep learning-based segmentation models, such as FCN [32], DeepLabv3+ [33], and Mask R-CNN [34], are being developed for use in a variety of applications and environments. For the purpose of semantic vegetation segmentation and to calculate the vegetation index from GSV imagery in this research work, FCN [22] and U-Net [35] semantic segmentation models are used. Their selection was based on their high precision and excellent performance in medical imaging area. The results of the experiments demonstrate that deep learning-based segmentation models are effective at segmenting vegetation images using semantic attributes.

2.3.1. Fully Convolutional Network (FCN)

Fully Convolutional Network (FCN) [22] uses locally connected layers, such as up-sampling, pooling, and convolution, to achieve segmentation. The architecture does not include any dense layers in order to reduce the amount of time it takes to compute and the number of parameters it requires. A segmentation map uses two paths to obtain output: the first is a down-sampling road, which is used to collect semantic/contextual information, and the second is an up-sampling path, which is used to recover spatial features. The architecture of FCN is depicted in Figure 6. Fully convolutional network architecture (FCN) was presented by Long et al. [22] for robust segmentation by adopting fully convolutional layers in place of the last fully linked layers. This approach allows the network to generate a dense pixel-wise prediction as a result of the advancement. The combination of up-sampled outputs with high-resolution activation maps results in improved localisation performance, which is then passed to the convolution layers to produce the correct output. The performance of FCN motivated to employ it as an important component of the proposed approach.

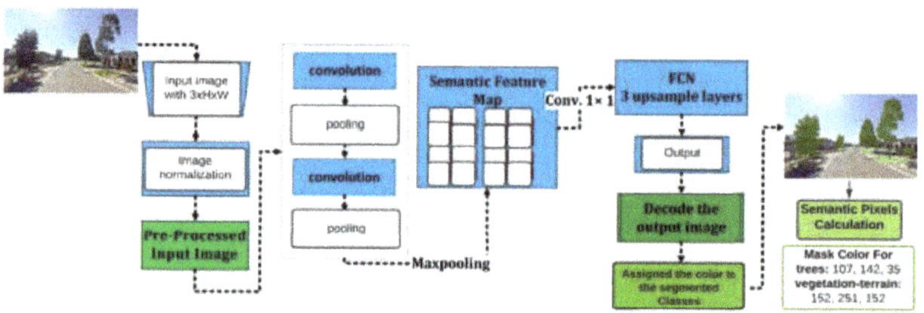

Figure 6. The architecture of fully convolution network (FCN) showing network processes. The masks for trees and vegetation are shown as RGB color codes.

2.3.2. U-Net

The second model employed in this work is U-Net [35], which has a similar encoder-decoder architecture to that of FCN but has two significant traits that distinguish it from the former. Since U-Net is symmetric, it bypasses the connections between the up-and down-sampling paths, which is useful when employed as a concatenation operator. Using the color variable, models assign a color to an item after they have been trained. The U-Net network (Figure 7) is built on an encoder-decoder architecture [35]. The encoder consists of a stack of convolutions and max-pooling layers that work together. The decoder is a symmetric expanding path that up-samples the feature maps with the use of learnable deconvolution filters, which can be learned. The major innovation brought about by this network is the way in which the so-called skip connections are utilised. To be more specific, they enable the concatenation of the output of the transposed convolution layers with the corresponding feature maps from the encoder stage during the convolution stage [36]. The main objective of this step is to get all the fine characteristics that were learned throughout the contracting stages in order to restore the spatial resolution of the original input image [35].

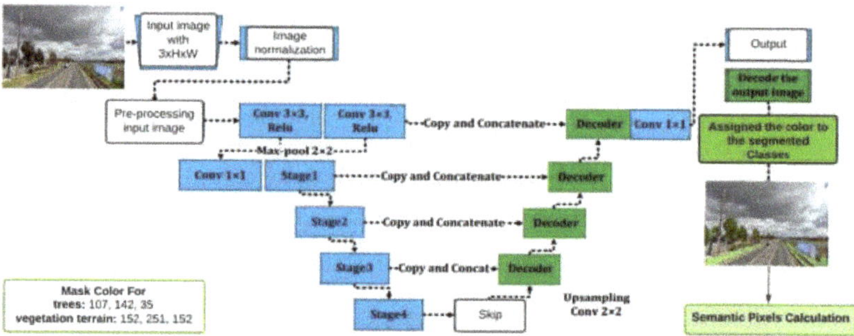

Figure 7. The architecture of U-Net showing network processes. The masks for trees and vegetation terrain are shown as RGB color codes.

According to standard practice, in the U-Net approach, the input image is initially processed by an encoder path, which is comprised of convolutional and pooling layers that degrade the spatial resolution of the input image. It is then followed by a decoder path that restores the original spatial imagery resolution by adopting up-sampling layers followed by convolutional layers, which is a technique known as "up-convolution". Apart from that, the network makes use of so-called skip connections, which connect the output of the relevant layers in the encoder path to the inputs of the decoder path by adding them to the inputs of the decoder path, whereas FCN allows pixel-wise classification performed for segmentation where features from initial convolutional layers are upsampled to develop deconvolution layers. These deconvolution layers develop the same size image, which is segmented on the basis of learnt features. Fine-tuning was performed to allow the network to learn efficient features of the vegetation region.

2.4. Vegetation Index Calculation from RGB Images

Various approaches are adopted in the literature for vegetation index calculation. Some of those are listed in Section 2.4.1. However, most of them used either color, threshold, or green area segmentation that might lead to promising results. To achieve robustness in vegetation index calculation, a semantic approach based on the unique color for each class of plants is proposed in this article. RGB color codes (107, 142, 35 and 152, 251, 152) for trees and vegetation terrain were assigned, respectively. After segmentation of the vegetation (trees and vegetation terrain), the respective masks are applied to calculate an accurate vegetation index as discussed in Section 2.4.2. For a better understanding of the RGB color space, the 3D data distribution in the RGB domain in Figure 8 is shown.

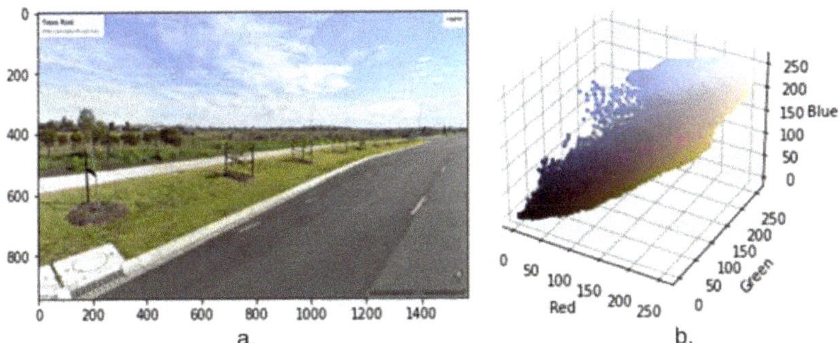

Figure 8. A sample image is presented in 3D color spaces for better understanding of data distribution: (**a**) sample image and (**b**) data distribution in RGB color space. As data in different color channels is tightly correlated, it provides inherent difficulties to differentiate color and semantic information in RGB domain.

2.4.1. Green View Index (GVI)

Mohamed et al. [37] explored extracting green vegetation from remotely sensed multispectral images. It has been identified that both, i.e., near-infrared and red bands, are being utilised quite often for vegetation detection. One of the primary reasons is that on red bands, the vegetation shows less absorption, and on infrared, they show great reflection. However, GSV images cover only the blue, red, green, and near-infrared bands. It was established by Yang et al. [12] that the GVI value was affected by two factors: the size of a tree's crown and the distance between the camera and the subject. A non-supervised classification methodology was used by Li et al. [13] to extract green vegetation from GSV images, which was justified by the fact that a significant number of GSV images were not available in the near-infrared band. According to their findings, green vegetation is significantly less reflective in red and blue bands. The red bands, on the other hand, are extremely reflective. As a result of this phenomenon, they developed extracting green vegetation from GSV images based on the natural hues of the images. There are a number of steps involved in the workflow.

Step-1: First of all, the subtraction of red band from green band generates Diff 1, and subtraction of blue band from green band gives Diff 2.

Step-2: Then the two images Diff 1 and Diff 2 were multiplied to create one Diff image. Normally, the green vegetation has greater reflectance values in the green band than the other two red and blue bands, and hence, the Diff image has positive green vegetation pixels.

Step-3: The pixels that have lower values in the green band as compared to the red and blue bands exhibit negative values in the Diff image

Step-4: As a result, an additional criterion was added stating that pixel values in the green band must be greater than those in the red band.

Usually, there were multiple spark points in the resulting images, after the initial classification images utilising the pixel-based classification approach were obtained as described in the steps above (Steps 1–4) [38]. The spectral vegetation variation has led to classifying individual pixels differently from their surrounding areas, leading to sparks in the classed image.

$$Green\ View\ Index = \frac{Number\ of\ green\ pixels\ segmented}{Total\ Number\ of\ pixels\ in\ an\ image} \tag{1}$$

The above equation gives information regarding the available greenery in the image. Yang et al. [12] proposed the Green View Index (GVI), which measures the visibility of urban woods in terms of greenery. Its GVI was defined as the relationship between the total green space and four image(s) taken at the intersection of the road and the sum of the four images taken at the intersection as shown in the following equation:

$$Green\ View\ Index = \frac{\sum_{i=1}^{4} Area_{g\_i}}{\sum_{i=1}^{4} Area_{t\_i}} * 100\% \tag{2}$$

where the $Area_{g\_i}$ presents the green pixels of the images taken in the direction of ith out of the four images taken in the (north, east, south, and west) directions. $Area_{t\_i}$ represents the total number of pixels in the image in the direction of ith. According to Li et al. [13], in this scenario, some surrounding vegetation may be missed from the calculation of the GVI since only four images cannot be seen in the fields of vision from the pedestrian view. Therefore, they modified the Equation (2) as below:

$$Green\ View\ Index = \frac{\sum_{i=1}^{6}\sum_{j=1}^{3} Area_{g\_ij}}{\sum_{i=1}^{6}\sum_{j=1}^{3} Area_{t\_ij}} * 100\% \tag{3}$$

where $Area_{g\_ij}$ denotes the number of green pixels in one of these images, which were taken in six directions with three vertical view angles for each sample site and were then averaged over all six directions. As a result, $Area_{t\_ij}$ represents the total amount of pixels included within each one of the eighteen GSV images.

2.4.2. The Proposed Semantic Vegetation Index (SVI)

For robust calculation of the vegetation index of each sample location on the road or street, the approach of semantic pixels (SP) is used, which is based on the unique color pixels assigned to vegetation's specific class (Vegetation terrain and trees) and are extracted based on the deep features through the use of a deep neural network. For index calculations, Google street view (GSV) images were used as such dataset is readily available. Therefore, in this investigation, a single image was used to calculate the vegetation index accurately based on the semantic pixels, so to cover all the vegetation area in the image. Hence, in each sample image, the number of semantic pixels will be determined as SP_a, with the area being the total semantic pixel numbers in one GSV image. The original Equation (1) has been updated and is now referred to as the semantic vegetation index (SVI).

$$SVI = \frac{\sum_{i=1}^{n} SP_{a\_i}}{\sum_{i=1}^{n} Area_{t\_i}} * 100\%$$ (4)

where SVI stands for semantic vegetation index, n is the total number of images, SP_{a_i} denotes the amount of semantic pixel area representing greenery in an image, and $Area_{t\_i}$ denotes the total amount of pixels in an image.

Similarly, to calculate the multiview semantic vegetation index, a total of six images covering the 360° horizontal environment with three vertical angles of, i.e., 45°, 0°, and −45° are used. The process is shown in Figure 5b to calculate the vegetation index accurately based on the semantic pixels so that to cover all vegetation area. Hence, in each sample site, the number of semantic pixels will be determined as SP_{a_ij}, with the Area being the total semantic pixel numbers in one of the 18 GSV images. Equation (3) has been modified to utilise semantic pixels for calculating the multiview semantic vegetation index ($MSVI$).

$$MSVI = \frac{\sum_{i=1}^{6}\sum_{j=1}^{3} SP_{a\_ij}}{\sum_{i=1}^{6}\sum_{j=1}^{3} Area_{t\_ij}} * 100\%$$ (5)

where $MSVI$ stands for multiview semantic vegetation index, SP_{a_ij} presents semantic pixels area of vegetation in input images which are taken from different pitch angles (45°, 0° and −45°) vertically as well as six horizontal direction covering 360° area, and $Area_{t\_ij}$ represents the sum of pixels in an image from the eighteen images of GSV.

3. Results

3.1. Preparation and Annotation of Data Set

For the experiments and implementation of the proposed model, first, a total of 3000 Google street view (GSV) images were downloaded using a python script. The next step was the pre-processing of the dataset so that the images could be used for training and testing phases. For the annotation of the training data, a cloud-based tool known as "Apeer", a ZEISS initiative [39], has been used. Image annotation generates labels that serve as the basis for machine learning training. Machine learning accuracy is determined by the amount of training data as well as the accuracy of annotations. The process of Annotation is summarised in Figure 9.

3.2. Experimental Environment Configuration

For the experiments and results, the hardware and software resources used are listed in Table 1.

Table 1. Configuration of experimental environment.

Item Name	Parameter
Central processing unit (CPU)	Intel i7 9700k
Operating system	MS Windows 10
Operating volatile memory	32GB RAM
Graphic processing unit (GPU)	Nvidia Titan RTX
Development environment configuration	Python 3.8 + TensorFlow 2.5 + CUDA 11.2 + cuDNN V8.1.0 + Visual Studio 2019

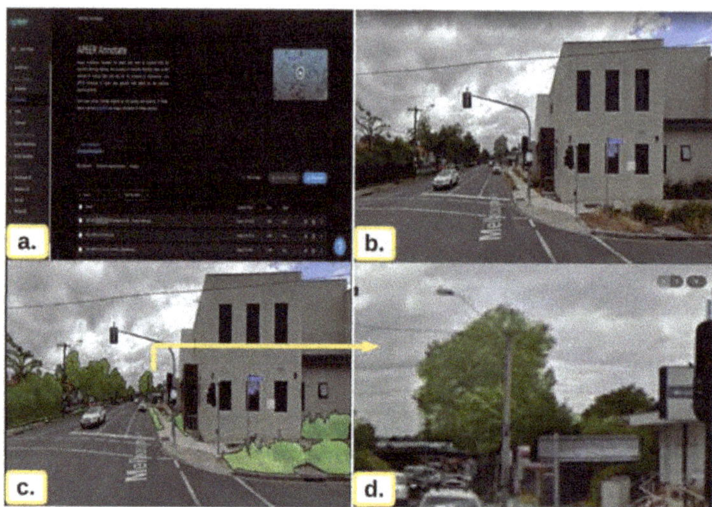

Figure 9. The process of data annotation shown in this figure: (**a**) a data annotation cloud based platform known as "Apeer", (**b**) sample image for annotation, (**c**) after completion of annotation, and (**d**) area zoomed for annotation in (**c**) and pointed with arrow.

3.3. Training of Deep Semantic Segmentation Models

The complete data set was split up into three distinct sections: training, validation, and testing sets, each comprising 80%, 15%, and 5% of the total, respectively. Before starting the training, hyperparameters were set to avoid the overfitting and underfitting issues of the model. The hyper-parameters used for the training of semantic segmentation model were the following: batch size kept as 16, learning rate as 0.0001, loss function as categorical cross-entropy, number of iteration/epochs as 200, NMS threshold as 0.45, and an optimiser as stochastic gradient descent "SGD". The training loss, validation loss, training accuracy and validation accuracy curve graphs are presented in Figures 10a,b and 11a,b for the FCN Model and the U-Net Model, respectively. The accuracy curve for the U-Net beats the accuracy curve for the FCN, as shown in the graph in Figure 11.

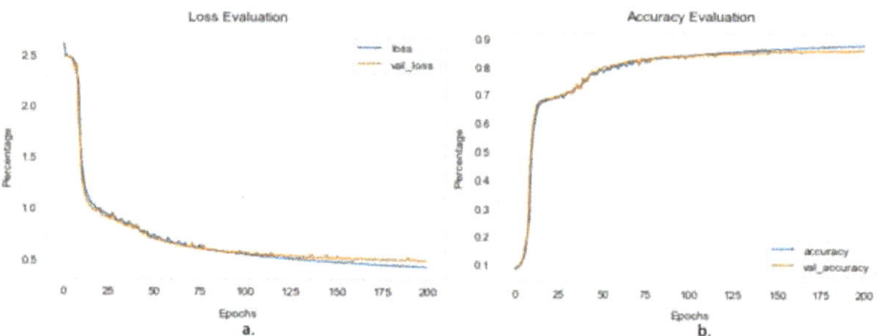

Figure 10. FCN segmentation model trend graphs for (**a**) training and validation loss and (**b**) training and validation accuracy.

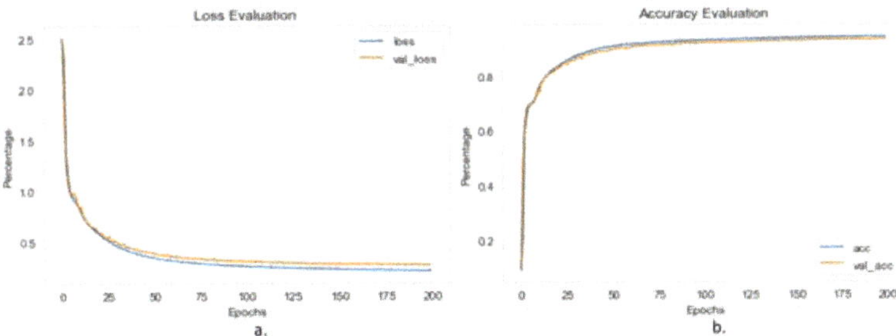

Figure 11. The U-Net segmentation model trend graphs for (**a**) training, validation loss and (**b**) training and validation accuracy.

Some of the sample results using FCN and U-Net segmentation models are shown in Figure 12, and vegetation index values are computed using Equation (4). The vegetation index values calculated from FCN for the test input images are 43%, 30% and 32%, while vegetation index values calculated from U-Net for the test input images are 41.4%, 33%, and 37%. The results show that the U-Net segmentation model gives comparatively more accurate and promising results than the FCN segmentation model. The ground truth results are computed manually to compare the results after masking manually and then calculating the pixel values of the vegetation, using Adobe Photoshop application software. The computed results are in percentage, as evident from Equation (4). Thus, on the basis of the ground truth data, U-Net vegetation index results are quite promising and are more close to the ground truth results.

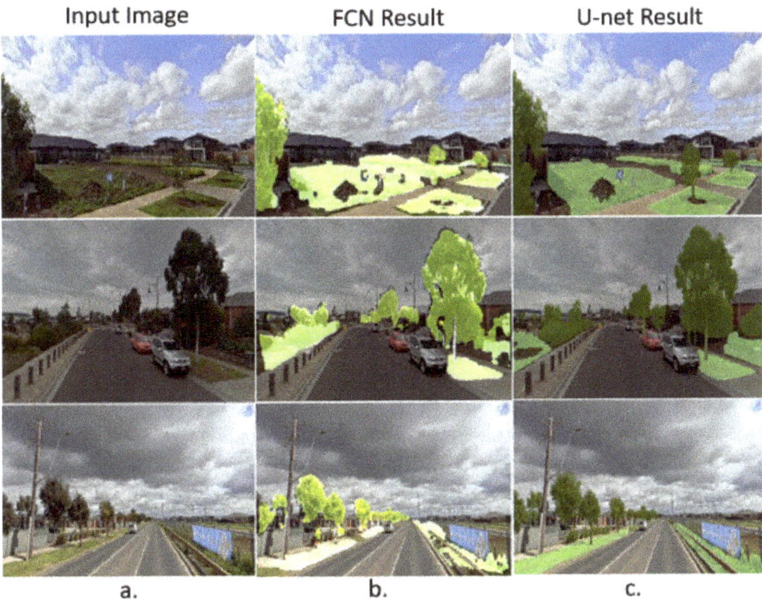

Figure 12. Segmentation and extraction of vegetation results from test input images: (**a**) input images, (**b**) results generated using FCN and (**c**) results generated using the U-Net model.

3.4. Performance Evaluation of Semantic Segmentation Networks

The performance of the semantic segmentation technique is evaluated using the metrics of precision, recall, F1-score, pixel accuracy (PA), intersection over union (IoU), and mean intersection over union (mIoU). Figure 13 shows the results of FCN and U-Net.

The accuracy of object contour segmentation is measured using the PA method, while the accuracy of an object detector on a particular dataset is measured using the IoU metric. The mIoU is the average of IoU and is defined to show the overall enhancement of semantic segmentation accuracy.

3.4.1. Precision, Recall, and F1-Score

FCN and U-Net segmentation models were compared in terms of precision, recall, and F-measure. The results of the comparison are shown in Table 2.

Precision is defined as the relationship between the number of accurately segmented vegetation pixels and the total number of pixels segregated as a vegetation region by the technique. The recall is the ratio between the number of successfully segmented vegetation pixels and the total number of vegetation pixels in the labelled image.

$$Precision = \frac{tp}{tp + fp} \tag{6}$$

$$Recall = \frac{tp}{tp + fn} \tag{7}$$

The equation of *F1-score* is shown below,

$$F1 - score = \frac{2 * Precision * Recall}{Precision + Recall} \tag{8}$$

3.4.2. Pixel Accuracy (PA)

In the evaluation of segmentation models, the pixel accuracy metric is the most commonly employed. It is defined as the accuracy of the pixel-wise prediction, given as

$$PA = \frac{\sum_{i=0}^{k} p_{ii}}{\sum_{i=0}^{k} \sum_{j=0}^{k} p_{ij}} \tag{9}$$

where k represents the total number of pixels in a test image, and p_{ii} is used to present the true positive predicted pixels as of class i, while p_{ij} presents the ground class i pixels as the number of pixels of class j.

3.4.3. Intersection Over Union (IoU)

Intersection over Union (IoU) is also known as the Jaccard Index [40], and it is a typically used assessment statistic for segmentation models that is used to calculate their overall performance. As shown below, it is commonly defined as the ratio of intersection and union areas between the projected segmentation map and ground truth.

$$IoU = \frac{p_{ii}}{\sum_{j=0}^{k} p_{ij} + \sum_{j=0}^{k} p_{ji} - p_{ii}} \tag{10}$$

where k indicates the total number of classes, p_{ii} represents the number of true positives, and p_{ij} and p_{ji} represent the number of false positives and false negatives, respectively.

3.4.4. Mean-IoU (mIoU)

mIoU is yet another matrix that is commonly used in segmentation models. It is calculated as the average value of all IoU label classes taken as a whole. This type of report is commonly used to summarise the performance of segmentation models, given as

$$mIoU = \frac{1}{k+1} \sum_{i=0}^{k} \frac{p_{ii}}{\sum_{j=0}^{k} p_{ij} + \sum_{j=0}^{k} p_{ji} - p_{ii}} \tag{11}$$

where k indicates the total number of classes, p_{ii} represents the number of true positives, and p_{ij} and p_{ji} represent the number of false positives and false negatives, respectively.

Figure 13 and Table 2 show the results achieved by different segmentation models used for vegetation index calculation on the basis of semantic pixels in an image. The U-Net model showed really promising results.

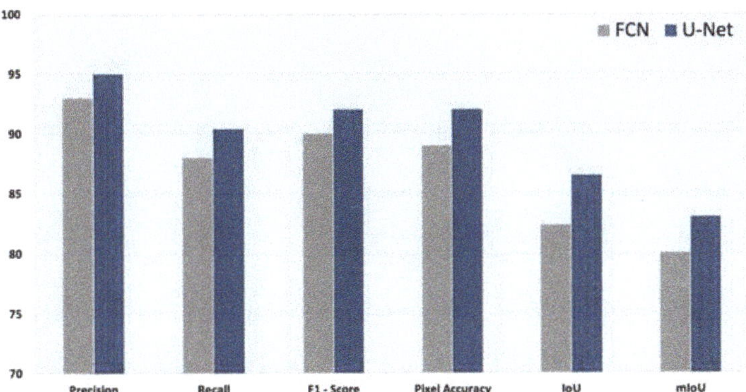

Figure 13. Performance evaluation of FCN and U-Net segmentation models.

Table 2. Performance evaluation results.

Segmentation Model	Precision	Recall	F1-Score	Pixel Accuracy	IoU	mIoU
FCN	93.2	87.3	90.1	89.4	82.3	80
U-Net	95	90.8	92.3	92.4	86.5	83

4. Comparative Analysis

The extraction of green vegetation from street view images is a difficult task because of a variety of factors, including the presence of shadows and spectral confusion between vegetation and other artificial green features (green walls, windows, shadows, signboards, etc.) Two studies are most relevant to this research: Yang et al. [12] used four GSV images in their work. As a result, Li et al. [13] modified the Green View Index (GVI) calculation, and they subsequently conducted a case study assessment of street vegetation using GSV images in the East Village of Manhattan District, New York City. They assert that the modified GVI may be a relatively objective measure of street-level greenery and that the use of GSV in conjunction with the modified GVI may be particularly effective in directing urban landscape planning and management practices.

For the purpose of comparison with the literature, sample images containing green vegetation, as well as green walls, signboards, and décor, were segmented and extracted for vegetation index calculation. Sample images segmentation results based on Li et al. [13] and Rencai et al. [15] are presented in Figure 14 and Table 3. From the results, it can be seen that the results of segmentation also included other green objects as vegetation because both the studies are principally based on green color. Both of the studies have mentioned this drawback in their studies and results, thus yielding an inaccurate vegetation index because of the inclusion of other green color objects. Hence, the vegetation index calculated values are higher as compared to our results. However, this study's results included vegetation only while ignoring other green color objects for calculating the index because it is based on semantic segmentation, thus giving an accurate vegetation index value.

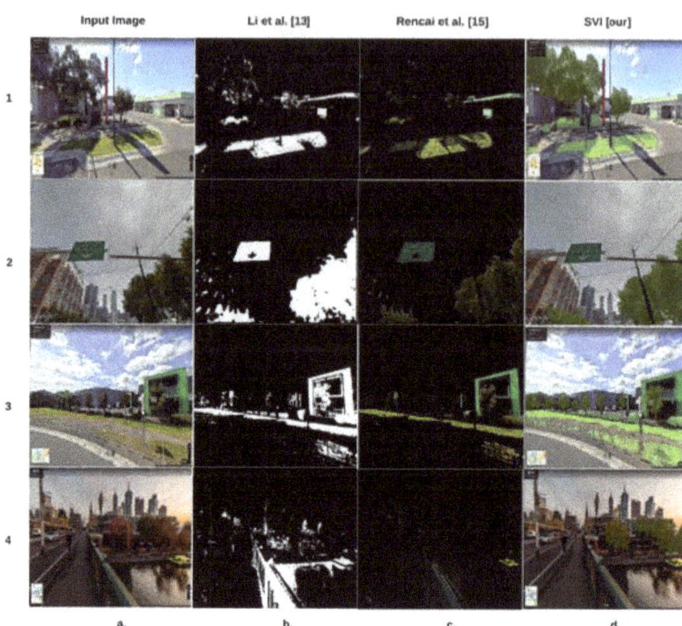

Figure 14. A sample of images and their segmentation (vegetation extraction) results using different approaches: (**a**) Sample input images, (**b**) Li et al. [13], and (**c**) Rencai et al. [15] and (**d**) SVI (proposed).

Table 3. Comparison table for vegetation segmentation and their vegetation index calculation using various vegetation extraction and index calculation approaches.

S.No.	GVI [13]	GVI [15]	SVI [Ours]
1	57.50%	55.91%	47.55%
2	46.62%	43.12%	35.44%
3	52.68%	51.25%	40.33%
4	43.08%	40.55%	27.42%

The multiview semantic vegetation index calculation for panoramic images taken at different angles horizontally (a) and with varying angles of pitch vertically (b), as shown in Figure 5, and the respective calculated vegetation index values are presented in Table 4. In the table, it is clear that the results of Li et al. [13] and Rencai et al. [15] are quite similar as both studies rely on the green color; hence, there are chances that during the calculation of the vegetation index, most of the time other objects of green color were included as mentioned before in the sample segmentation result shown in Table 3. Therefore, the results are inaccurate, and the vegetation index percentage indicated is larger than ours because both comparison studies employed the green area index, and the tram in the image was also used to compute green color in those studies shown in Figure 5. On the other hand, the proposed model extracted only vegetation index. The input image on the second row in the Figure 14 is taken from the paper by Li et al. [13] only for comparison purposes. There, they mentioned that their algorithm is based on the green color, thus including another green object during the calculation of the green view index.

Table 4. Comparative analysis of vegetation index calculation through various approaches.

Li et al. [13]	Rencai et al. [15]	MSVI [Proposed]
63.40%	62.9%	56.19%

5. Discussion

Based on the research study, the semantic segmentation leads to accurate index calculation. The publicly available GSV imagery of the urban areas was used to quantify street greenery, i.e., SVI of the urban streets. GSV are freely available to the public and can be used in machine learning/computer vision in an efficient way to perform multiple activities automatically. The SVI can be utilised as useful information/data for a better assessment of urban greenery by considering people's envisioned vegetation on a street scale for urban planners and others. To assess the greenery of street vegetation, GSV images captured from the ground should be similar to those of pedestrians.

A single vertical point of view is insufficient to express correctly the surrounding vegetation index that pedestrians may observe; two vertical points of view are required. Therefore, the multiview semantic vegetation index (MSVI) is employed for six GSV images in this experiment to calculate the vegetation index, each spanning a 360° horizontal and three vertical angles of 45°, 0° and −45°, to calculate the vegetation index appropriately on the basis of the semantic pixels.

According to the findings of this study, GSV images are qualified for assessing street greenery, and the modified GVI may be a more objective measurement of street-level greenery. The multiview semantic vegetation index (MSVI) took advantage of the characteristics of GSV images, used 18 GSV images taken from different viewing angles, making the index more efficient for evaluating street greenery in urban areas. Because it measures the amount of visible urban greenery on the ground, the SVI formula is simpler to understand for the general public. As a result, it can give a monitoring tool to analyze gains or losses in urban vegetation. It may serve to help urban planners select the sites, sizes and varieties of greenery for best effect in the planning stage of an urban greening program. It, therefore, seems to be a promising instrument, not a mere gadget for users, for future urban planning and urban environmental management.

The strength of SVI lies in its robustness to color variations and viewpoint constraints. The limitation of the approach is its reliance on captured viewpoints and attributes of the captured image, like its zoom level and image quality. Therefore, if SVI is utilised for long-term vegetation monitoring, it is proposed that proper dataset normalisation and image registration scale or affine invariant [41] be used before SVI estimation.

6. Conclusions

This research paper proposes a robust vegetation index based on semantic segmentation called a multiview semantic vegetation index (MSVI). The Google Street View (GSV) imagery dataset is used for calculating and indexing the vegetation cover of an urban area of the Wyndham City Council in Melbourne, Australia. The MSVI is based on the deep features learned from a deep neural network to calculate the vegetation index of each sample location in the urban area. For vegetation segmentation, different deep learning-based semantic segmentation models, such as FCN and U-Net, were tried. Using the GSV data set, both segmentation models were trained and tested to improve their overall performance.

The proposed method for segmenting urban vegetation areas has yielded promising results. Generally speaking, U-Net shows better results than FCN. FCN and U-Net models achieve *Precision* of 93.2% and 95%, *Recall* of 87.3% and 90.8%, *F1-score* of 90.1% and 92.3%, pixel accuracy *(PA)* of 89.4% and 92.4%, *IoU* of 82.3% and 86.5%, and *mIoU* of 80% and 83%, respectively. The proposed MSVI index measures the broad visible urban greenery on the ground, which can assist urban planners and strategists in better understanding urban green spaces.

We intend to use this approach in the future for real-time vegetation index calculation using Google panoramic cameras such as Pilot Era 360°, Insta360 pro, and Insta360 pro2, which will be of great help in the quest for ecological improvement.

Author Contributions: Conceptualization, A.K.; methodology, A.K.; software, A.K.; validation, A.K. and W.A; formal analysis, A.K.; investigation, A.K.; resources, A.K.; data curation, A.K. and W.A.; writing—original draft preparation, A.K.; writing—review and editing, W.A., A.K, A.U. and R.W.R.; visualization, A.K. and W.A.; supervision, A.U. and R.W.R.; project administration, A.K., A.U. and R.W.R.; funding acquisition, A.U. and R.W.R. All authors have read and agreed to the published version of the manuscript.

Funding: This study received no external funding. However, Victoria University, Footscray 3011, Australia and Charles Sturt University, Port Macquarie (Campus), NSW 2444, Australia, equally funded.

Institutional Review Board Statement: Not applicable.

Informed Consent Statement: Not applicable.

Data Availability Statement: The data used to support the findings of this study are available subject to approval from the relevant departments through the corresponding author upon request.

Conflicts of Interest: The authors declare no conflict of interest.

References

1. Song, X.P.; Hansen, M.C.; Stehman, S.V.; Potapov, P.V.; Tyukavina, A.; Vermote, E.F.; Townshend, J.R. Global land change from 1982 to 2016. *Nature* **2018**, *560*, 639–643. [CrossRef] [PubMed]
2. Edgeworth, M.; Ellis, E.C.; Gibbard, P.; Neal, C.; Ellis, M. The chronostratigraphic method is unsuitable for determining the start of the Anthropocene. *Prog. Phys. Geogr.* **2019**, *43*, 334–344. [CrossRef]
3. Rosan, T.M.; Aragão, L.E.; Oliveras, I.; Phillips, O.L.; Malhi, Y.; Gloor, E.; Wagner, F.H. Extensive 21st-Century Woody Encroachment in South America's Savanna. *Geophys. Res. Lett.* **2019**, *46*, 6594–6603. [CrossRef]
4. Wolf, K.L. Business district streetscapes, trees, and consumer response. *J. For.* **2005**, *103*, 396–400.
5. Appleyard, D. Urban trees, urban forests: What do they mean. In Proceedings of the National Urban Forestry Conference, Washington, DC, USA, 13–16 November 1979; pp. 138–155.
6. Nowak, D.J.; Hoehn, R.; Crane, D.E. Oxygen production by urban trees in the United States. *Arboric. Urban For.* **2007**, *33*, 220–226.
7. Chen, X.L.; Zhao, H.M.; Li, P.X.; Yin, Z.Y. Remote sensing image-based analysis of the relationship between urban heat island and land use/cover changes. *Remote Sens. Environ.* **2006**, *104*, 133–146. [CrossRef]
8. Onishi, A.; Cao, X.; Ito, T.; Shi, F.; Imura, H. Evaluating the potential for urban heat-island mitigation by greening parking lots. *Urban For. Urban Green.* **2010**, *9*, 323–332. [CrossRef]
9. Camacho-Cervantes, M.; Schondube, J.E.; Castillo, A.; MacGregor-Fors, I. How do people perceive urban trees? Assessing likes and dislikes in relation to the trees of a city. *Urban Ecosyst.* **2014**, *17*, 761–773. [CrossRef]
10. Balram, S.; Dragićević, S. Attitudes toward urban green spaces: Integrating questionnaire survey and collaborative GIS techniques to improve attitude measurements. *Landsc. Urban Plan.* **2005**, *71*, 147–162. [CrossRef]
11. Gao, L.; Wang, X.; Johnson, B.A.; Tian, Q.; Wang, Y.; Verrelst, J.; Mu, X.; Gu, X. Remote sensing algorithms for estimation of fractional vegetation cover using pure vegetation index values: A review. *ISPRS J. Photogramm. Remote Sens.* **2020**, *159*, 364–377. [CrossRef]
12. Yang, J.; Zhao, L.; Mcbride, J.; Gong, P. Can you see green? Assessing the visibility of urban forests in cities. *Landsc. Urban Plan.* **2009**, *91*, 97–104. [CrossRef]
13. Li, X.; Zhang, C.; Li, W.; Ricard, R.; Meng, Q.; Zhang, W. Assessing street-level urban greenery using Google Street View and a modified green view index. *Urban For. Urban Green.* **2015**, *14*, 675–685. [CrossRef]
14. Li, X.; Zhang, C.; Li, W.; Kuzovkina, Y.A. Environmental inequities in terms of different types of urban greenery in Hartford, Connecticut. *Urban For. Urban Green.* **2016**, *18*, 163–172. [CrossRef]
15. Dong, R.; Zhang, Y.; Zhao, J. How green are the streets within the sixth ring road of Beijing? An analysis based on tencent street view pictures and the green view index. *Int. J. Environ. Res. Public Health* **2018**, *15*, 1367. [CrossRef] [PubMed]
16. Zhang, Y.; Dong, R. Impacts of street-visible greenery on housing prices: Evidence from a hedonic price model and a massive street view image dataset in Beijing. *ISPRS Int. J. Geo Inf.* **2018**, *7*, 104. [CrossRef]
17. Long, Y.; Liu, L. How green are the streets? An analysis for central areas of Chinese cities using Tencent Street View. *PLoS ONE* **2017**, *12*, e0171110. [CrossRef]
18. Cheng, L.; Chu, S.; Zong, W.; Li, S.; Wu, J.; Li, M. Use of tencent street view imagery for visual perception of streets. *ISPRS Int. J. Geo Inf.* **2017**, *6*, 265. [CrossRef]
19. Kendal, D.; Hauser, C.E.; Garrard, G.E.; Jellinek, S.; Giljohann, K.M.; Moore, J.L. Quantifying plant colour and colour difference as perceived by humans using digital images. *PLoS ONE* **2013**, *8*, e72296.
20. Lopatin, J.; Dolos, K.; Kattenborn, T.; Fassnacht, F.E. How canopy shadow affects invasive plant species classification in high spatial resolution remote sensing. *Remote Sens. Ecol. Conserv.* **2019**, *5*, 302–317. [CrossRef]
21. Schiefer, F.; Kattenborn, T.; Frick, A.; Frey, J.; Schall, P.; Koch, B.; Schmidtlein, S. Mapping forest tree species in high resolution UAV-based RGB-imagery by means of convolutional neural networks. *ISPRS J. Photogramm. Remote Sens.* **2020**, *170*, 205–215.

22. Long, J.; Shelhamer, E.; Darrell, T. Fully convolutional networks for semantic segmentation. In Proceedings of the IEEE Conference on Computer Vision and Pattern Recognition, Boston, MA, USA, 7–12 June 2015; pp. 3431–3440.
23. Dvornik, N.; Shmelkov, K.; Mairal, J.; Schmid, C. Blitznet: A real-time deep network for scene understanding. In Proceedings of the IEEE International Conference on Computer Vision, Venice, Italy, 22–29 October 2017; pp. 4154–4162.
24. Li, Y.; Qi, H.; Dai, J.; Ji, X.; Wei, Y. Fully convolutional instance-aware semantic segmentation. In Proceedings of the IEEE Conference on Computer Vision and Pattern Recognition, Honolulu, HI, USA, 21–26 July 2017; pp. 2359–2367.
25. Chen, L.C.; Yang, Y.; Wang, J.; Xu, W.; Yuille, A.L. Attention to scale: Scale-aware semantic image segmentation. In Proceedings of the IEEE Conference on Computer Vision and Pattern Recognition, Las Vegas, NV, USA, 27–30 June 2016; pp. 3640–3649.
26. Council, W.C. Street Tree Planting | Wyndham City. 2021. Available online: https://www.wyndham.vic.gov.au/treeplanting (accessed on 15 August 2021).
27. Street View Static API Overview | Google Developers. Available online: https://developers.google.com/maps/documentation/streetview/overview (accessed on 17 August 2021).
28. Tsai, V.J.; Chang, C.T. Three-dimensional positioning from Google street view panoramas. *IET Image Process.* **2013**, *7*, 229–239. [CrossRef]
29. Hao, S.; Zhou, Y.; Guo, Y. A brief survey on semantic segmentation with deep learning. *Neurocomputing* **2020**, *406*, 302–321. [CrossRef]
30. Uhrig, J.; Cordts, M.; Franke, U.; Brox, T. Pixel-level encoding and depth layering for instance-level semantic labeling. In *German Conference on Pattern Recognition*; Springer: Berlin/Heidelberg, Germany, 2016; pp. 14–25.
31. Guo, Y.; Liu, Y.; Georgiou, T.; Lew, M.S. A review of semantic segmentation using deep neural networks. *Int. J. Multimed. Inf. Retr.* **2018**, *7*, 87–93. [CrossRef]
32. Liu, X.; Deng, Z.; Yang, Y. Recent progress in semantic image segmentation. *Artif. Intell. Rev.* **2019**, *52*, 1089–1106. [CrossRef]
33. Chen, L.C.; Zhu, Y.; Papandreou, G.; Schroff, F.; Adam, H. Encoder-decoder with atrous separable convolution for semantic image segmentation. In Proceedings of the European Conference on Computer Vision (ECCV), Munich, Germany, 8–14 September 2018; pp. 801–818.
34. He, K.; Gkioxari, G.; Dollár, P.; Girshick, R. Mask r-cnn. In Proceedings of the IEEE International Conference on Computer Vision, Venice, Italy, 22–29 October 2017; pp. 2961–2969.
35. Ronneberger, O.; Fischer, P.; Brox, T. U-net: Convolutional networks for biomedical image segmentation. In *Proceedings of the International Conference on Medical Image Computing and Computer-Assisted Intervention—MICCAI 2015*; Lecture Notes in Computer Science; Springer: Cham, Switzerland, 2015; Volume 9351, pp. 234–241. [CrossRef]
36. Garcia-Garcia, A.; Orts-Escolano, S.; Oprea, S.; Villena-Martinez, V.; Garcia-Rodriguez, J. A review on deep learning techniques applied to semantic segmentation. *arXiv* **2017**, arXiv:1704.06857.
37. Almeer, M.H. Vegetation extraction from free google earth images of deserts using a robust BPNN approach in HSV Space. *Int. J. Adv. Res. Comput. Commun. Eng.* **2012**, *1*, 134–140.
38. Blaschke, T.; Lang, S.; Lorup, E.; Strobl, J.; Zeil, P. Object-oriented image processing in an integrated GIS/remote sensing environment and perspectives for environmental applications. *Environ. Inf. Plan. Politics Public* **2000**, *2*, 555–570.
39. APEER. Available online: https://www.apeer.com/ (accessed on 15 August 2021).
40. Hamers, L. Similarity measures in scientometric research: The Jaccard index versus Salton's cosine formula. *Inf. Process. Manag.* **1989**, *25*, 315–318. [CrossRef]
41. Khan, A.; Ulhaq, A.; Robinson, R.W. Multi-temporal registration of environmental imagery using affine invariant convolutional features. In *Pacific-Rim Symposium on Image and Video Technology*; Springer: Berlin/Heidelberg, Germany, 2019; pp. 269–280.

 **MDPI**

Article

Fine-Grained Tidal Flat Waterbody Extraction Method (FYOLOv3) for High-Resolution Remote Sensing Images

Lili Zhang [1], Yu Fan [1], Ruijie Yan [1], Yehong Shao [2], Gaoxu Wang [3,*] and Jisen Wu [1]

[1]   College of Computer and Information Engineering, Hohai University, Nanjing 211100, China; lilzhang@hhu.edu.cn (L.Z.); 191307020015@hhu.edu.cn (Y.F.); yanruijie@hhu.edu.cn (R.Y.); 171307030009@hhu.edu.cn (J.W.)
[2]   Arts and Sciences, Ohio University Southern, Ironton, OH 45638, USA; yehongshao@gmail.com
[3]   State Key Laboratory of Hydrology—Water Resources and Hydraulic Engineering, Nanjing Hydraulic Research Institute, Nanjing 210029, China
*   Correspondence: gxwang@nhri.cn

Citation: Zhang, L.; Fan, Y.; Yan, R.; Shao, Y.; Wang, G.; Wu, J. Fine-Grained Tidal Flat Waterbody Extraction Method (FYOLOv3) for High-Resolution Remote Sensing Images. *Remote Sens.* **2021**, 13, 2594. https://doi.org/10.3390/rs13132594

Academic Editors: Anwaar Ulhaq and Douglas Pinto Sampaio Gomes

Received: 21 May 2021
Accepted: 25 June 2021
Published: 2 July 2021

Publisher's Note: MDPI stays neutral with regard to jurisdictional claims in published maps and institutional affiliations.

Abstract: The tidal flat is long and narrow area along rivers and coasts with high sediment content, so there is little feature difference between the waterbody and the background, and the boundary of the waterbody is blurry. The existing waterbody extraction methods are mostly used for the extraction of large water bodies like rivers and lakes, whereas less attention has been paid to tidal flat waterbody extraction. Extracting tidal flat waterbody accurately from high-resolution remote sensing imagery is a great challenge. In order to solve the low accuracy problem of tidal flat waterbody extraction, we propose a fine-grained tidal flat waterbody extraction method, named FYOLOv3, which can extract tidal flat water with high accuracy. The FYOLOv3 mainly includes three parts: an improved object detection network based on YOLOv3 (Seattle, WA, USA), a fully convolutional network (FCN) without pooling layers, and a similarity algorithm for water extraction. The improved object detection network uses 13 convolutional layers instead of Darknet-53 as the model backbone network, which guarantees the water detection accuracy while reducing the time cost and alleviating the overfitting phenomenon; secondly, the FCN without pooling layers is proposed to obtain the accurate pixel value of the tidal flat waterbody by learning the semantic information; finally, a similarity algorithm for water extraction is proposed to distinguish the waterbody from non-water pixel by pixel to improve the extraction accuracy of tidal flat water bodies. Compared to the other convolutional neural network (CNN) models, the experiments show that our method has higher accuracy on the waterbody extraction of tidal flats from remote sensing images, and the *IoU* of our method is 2.43% higher than YOLOv3 and 3.7% higher than U-Net (Freiburg, Germany).

Keywords: tidal flat water; YOLOv3; similarity algorithm for water extraction

1. Introduction

Water resources are closely related to human survival and development, and many researchers focus on how to obtain water resource information quickly and accurately. The extraction and detection of the water bodies from remote sensing images is one of the main ways to obtain water resource information. It can be widely applied in ecosystem protection and restoration, river supervision, pollution control, and infrastructure construction [1,2]. In recent years, with the rapid development of remote sensing satellite technology, obtaining water resource information from remote sensing images [3] has gradually replaced manual measurement, and the images are widely applied in water resource surveys and flood predictions.

At present, scholars have proposed a variety of water extraction methods for different satellite imagery, which can be summarized into three categories: visual interpretation methods [4], extraction methods based on spectral bands [5–9], and machine learning methods [10–12]. However, these methods are mainly applied to extract large water bodies like rivers and lakes, and there are few waterbody extraction methods for tidal flats. The tidal flat area [13] refers to the tidal invasion area between the high tide level and the low tide level along rivers and coasts, etc. The water bodies in this kind of area are relatively long and narrow, with high sediment content. Due to the influence of tides, there is little feature difference between the waterbody and the background, and the boundary of the waterbody is blurry. Meanwhile, the mixture of water and sand makes the spectral band characteristics of the water in the tidal flat area different from the water in the other areas. Therefore, the methods based on spectral bands are not suitable for tidal flat waterbody extraction. The machine learning method used for water extraction is usually based on supervised learning, so the training dataset is necessary. However, there is not public training dataset for tidal flat waterbody extraction. Hence the machine learning methods usually have poor ability and do not learn effectively due to the limited training dataset and have an accuracy bottleneck in the water extraction as a result.

The boundary of the waterbody is blurry in tidal flat area, in order to solve the low accuracy problem of its waterbody extraction caused by little feature difference between the waterbody and the background, this paper proposes a fine-grained tidal flat waterbody extraction method, named FYOLOv3. The FYOLOv3 mainly includes three parts: an improved object detection network based on YOLOv3 (Seattle, WA, USA), a fully convolutional network (FCN) without pooling layers, and a similarity algorithm for water extraction.

In this paper, our contributions are as follows:

(1) An improved object detection network was introduced, which contains two modules, one is a 13-layer convolutional neural network (CNN) as the backbone network, and the other is the feature pyramid network for multi-scale water detection.

(2) A FCN without pooling layers is proposed to obtain the accurate pixel value of the tidal flat waterbody by learning the semantic information, complete the initial extraction of the waterbody and realize cross-channel information fusion.

(3) A similarity algorithm for water extraction is proposed to distinguish the waterbody from non-water pixel by pixel to improve the extraction accuracy of tidal flat waterbodies, in which a standard water pixel valued and similarity between the water pixels and the standard water pixels are introduced, respectively.

The rest of this paper is organized as follows. In Section 2, we introduce some classical methods and analyze the YOLO models. In Section 3, a fine-grained tidal flat waterbody extraction method FYOLOv3 is described in detail. The experiments and analysis are presented in Section 4. Finally, the conclusion of this paper with some discussions and future work are given in Section 5.

2. Related Work

2.1. Water Extraction Methods

Spectral band analysis methods are the earlier methods for waterbody extraction from remote sensing images [5–9] by analyzing the differences of absorption and reflection of different ground objects for each band spectrum, then obtaining the water region in the remote sensing images. There are three methods based on spectral analysis: single band threshold method [14], multi-band spectral relationship method [15], and water index method [8]. Xu et al. [8] proposed an improved normalized difference water index (MNDWI) based on the band combination of the normalized water index. The experiments show that the method is efficient for the extraction of urban water bodies, and effectively solves the influence of urban building shadow. Guo et al. [7] proposed a weighted normalized difference water index (WNDWI) to solve the influence of turbid water, small water bodies and shadow areas on water extraction. The method was tested on Landsat images and achieved good results. Methods based on spectral analysis usually only use the spectral information of remote sensing images, which does not effectively use the texture, space, surrounding background, and other information, so its extraction ability has certain limitations. These methods have specific requirements for the band of remote sensing images and have low applicability as a result.

Some machine learning methods, such as support vector machine (SVM) and maximum likelihood classification [10–12] try to balance the learning effectiveness and the interpretability of the models and provide a solution framework for the classification problem of limited samples. This kind of method improves the accuracy of target extraction in a certain range by learning the distribution characteristics of the training data. However, they have poor ability and do not learn effectively due to the limited training dataset and have an accuracy bottleneck in the water extraction.

With the concept of deep learning proposed by Hinton et al. [16] in 2006 and the outstanding achievements of deep convolution neural network proposed by Alex [17] in natural images recognition in 2012, deep learning ushered in a new research phase. Many experts and scholars began to apply deep learning technology to obtain object extraction from remote sensing images. Zhong et al. [18] used convolution neural network model to extract waterbody from remote sensing images, and the experiments showed that convolution neural network is more efficient to extract waterbody from remote sensing images than normalized water index. Liang et al. [19] introduced dense connection structure in the full convolution network to reduce the shallow feature loss, get more detailed information from the remote sensing images, and achieve better water extraction. Song et al. [20] used the self-learning ability of deep learning to construct a modified Mask R-CNN method which integrates bottom-up and top-down processes for water recognition. Yu et al. [21] presented a novel deep learning framework for waterbody extraction from Landsat images considering both its spectral and spatial information, which is a hybrid of CNN and logistic regression classifier. Li et al. [22] adopted a fully convolutional network (FCN) to extract water bodies in the case of limited training data, which consists of an encoder for extracting multiscale features and a decoder for recovering spatial contexts. Wang et al. [23] proposed an end-to-end trainable model named the multi-scale lake water extraction network (MSLWENet) to extract lake water from Google remote sensing images. Yu et al. [24] developed a novel self-attention capsule feature pyramid network (SA-CapsFPN) to extract water bodies from remote sensing images. Li et al. [25] built a deep learning model for water extraction based on the EfficientNet-B5 (Perdriel, Argentina).

2.2. YOLO Models

The excellent performance of deep convolution neural network [17] has been demonstrated in computer vision. Recently, YOLO models such as YOLOv1 (Seattle, WA, USA) [26], YOLOv2 [27], and YOLOv3 [28], were proposed one after another. The YOLOv1 model is based on GoogLeNet (Mountain View, CA, USA) [29], which is mainly composed of convolutional layers and fully connected layers to achieve the object detection fast. The model transforms the object detection problem into coordinate regression problem and carries out the classification and regression of target objects. Because the two prediction frames generated in YOLOv1 (Seattle, WA, USA) for each lattice in the images can only predict one target object, the detection accuracy of adjacent objects whose center point falls in the same lattice is reduced as a result. In view of the above shortcomings, YOLOv2 (Seattle, WA, USA) proposes a variety of strategies to improve the network framework, which significantly improves the speed and accuracy of object detection. In order to further optimize the YOLO models, the DarkNet-53 network is used as the object feature extractor in YOLOv3 (Seattle, WA, USA) model, and the output module uses the feature pyramid structure to achieve three-way outputs to complete the accurate detection of the targets with different sizes [28].

3. Methodology

To solve the low accuracy problem of water extraction for tidal flats, this paper proposes a fine-grained tidal flat waterbody extraction method for high-resolution remote sensing images, named FYOLOv3. The key parts of our method are as follows, firstly, the improved object detection network based on YOLOv3 (Seattle, WA, USA) is proposed and used to locate the tidal flat waterbody, and the frame coordinates of the corresponding waterbody are obtained; secondly, four images with size of 32 × 32 are clipped from the obtained border region, which are used as the input of the FCN without pooling layers to get the initial waterbody extraction; finally, the similarity algorithm for water extraction is used to judge all pixels in the obtained initial waterbody region to optimize and improve the initial waterbody extraction. We list the steps of our method as follows:

1. Construction of training dataset: This part mainly includes the data preprocessing, data augmentation and waterbody labeling of remote sensing images.
2. Model construction: Based on the YOLOv3 (Seattle, WA, USA) model, this paper constructs an improved network model for water detection. It uses 13 convolutional layers as the model backbone network to meet the accuracy requirement of water detection, while reducing the time cost and alleviating the overfitting phenomenon; it also uses two branch structures as the output module to avoid the problem of missing extraction in the waterbody extraction caused by the small prior box. FCN without pooling layers followed the improved object detection network to obtain the semantic information of waterbody in a tidal flat area.
3. Model training: The cross entropy is used as the loss function, and the backpropagation algorithm is used to train the internal parameters of the network model.
4. Detection and initial extraction: The trained network models are used to detect the water from the remote sensing images to locate the waterbody area and obtain the initial waterbody extraction, respectively.
5. Similarity algorithm for water extraction: This algorithm is used to optimize and improve the initial waterbody extraction by similarity judgment.

The architecture of our method is shown in Figure 1, where the 13-layer CNN is constructed for water feature extraction and mainly composed of convolutional layers, pooling layers, and batch standardization layers. The multi-scale feature pyramid network uses the different feature maps to get the narrow and long waterbodies and small waterbodies in the tidal flat area, respectively. Hence, our object detection network can guarantee the water detection accuracy while reducing the time cost and alleviating the overfitting phenomenon.

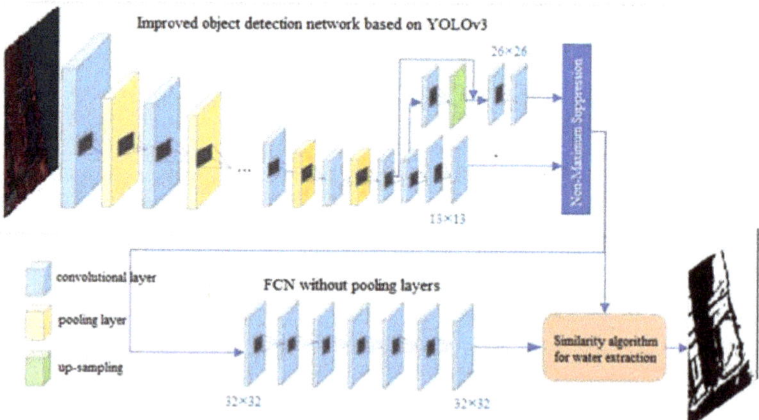

Figure 1. Fine-grained tidal flat waterbody extraction method.

3.1. Construction of Training Dataset

3.1.1. Preprocessing

The GF-2 remote sensing satellite is the first civil high-resolution satellite in China, and it was successfully launched in 2014. GF-2 satellite has high spatial resolution, accurate positioning, and strong maneuverability. The remote sensing images used in this paper are the Level-1 product data. Therefore, it is necessary to preprocess the remote sensing images first. The preprocessing of GF-2 remote sensing images used in this paper mainly includes radiometric calibration [30], atmospheric correction [31], orthorectification [32], and image fusion [33].

1. Radiometric correction and orthorectification of multispectral images

Radiometric correction includes two parts: radiometric calibration and atmospheric correction. Radiometric calibration refers to convert the brightness value of pixels into absolute radiance value, which helps researchers to compare remote sensing images acquired from different types of sensors at different times. Atmospheric correction refers to the process of eliminating the radiation error caused by atmospheric influence and obtaining the true reflectance of surface objects. Orthorectification is to correct the geometric distortion of remote sensing images and the plane orthophotos are obtained at last. The preprocessing example of the multispectral images is shown in Figure 2.

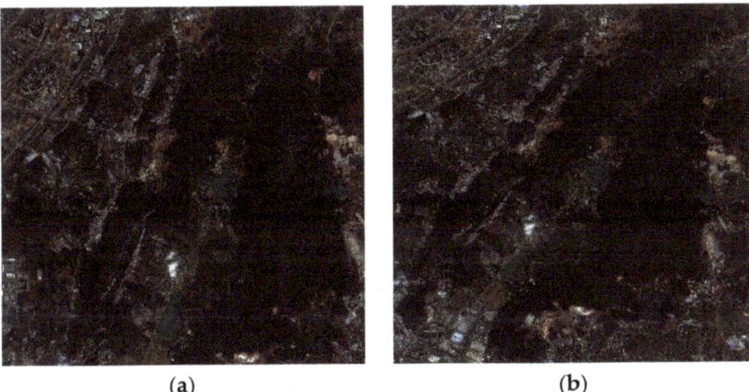

(a)   (b)

Figure 2. Comparison of multispectral images before and after preprocessing. (**a**) Multispectral image before preprocessing; (**b**) multispectral image after preprocessing.

2. Orthorectification of panchromatic images

Different from multispectral images, the band range of panchromatic images in GF-2 is 0.45–0.90 μm, which includes multiple wavelength ranges. The attenuation of the atmosphere is selective for the lights with different wavelengths, and each wavelength is affected by the atmosphere differently. Therefore, it is usually impossible to carry out an atmospheric correction for panchromatic images. The number and distribution of controlled points in a remote sensing image influence the error of the orthorectification and the mean square error is used to evaluate the accuracy of orthorectification. Fan et al. made the accuracy analysis of GF-2 satellite image according to the mentioned evaluation indexes [34], and the RPC orthorectification was proved better to correct the geometric distortion in panchromatic images. Hence, we use the RPC orthorectification to deal with the panchromatic images in this paper, and the experiment is shown in Figure 3.

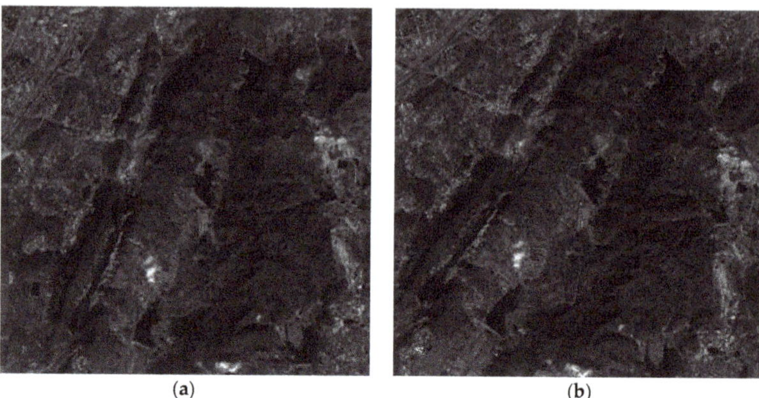

(a)   (b)

Figure 3. Comparison of panchromatic images before and after orthorectification. (**a**) Panchromatic image before orthorectification; (**b**) panchromatic image after orthorectification.

3. Image fusion

Image fusion is often used to enrich the image information. It fuses the images of the same area from different channels and finally obtains the fused images with more information and higher quality.

In this paper, the NNDiffuse Pan Sharpening [35] method is used to fuse the multi-spectral images and panchromatic images. The multispectral images and the panchromatic images are obtained synchronously by different sensors installed in the GF-2, and the former has higher resolution but less spectral information, and the latter has more spectral information and lower resolution. If we fuse them, we could get the fused image with high resolution and more spectral information, as shown in Figure 4.

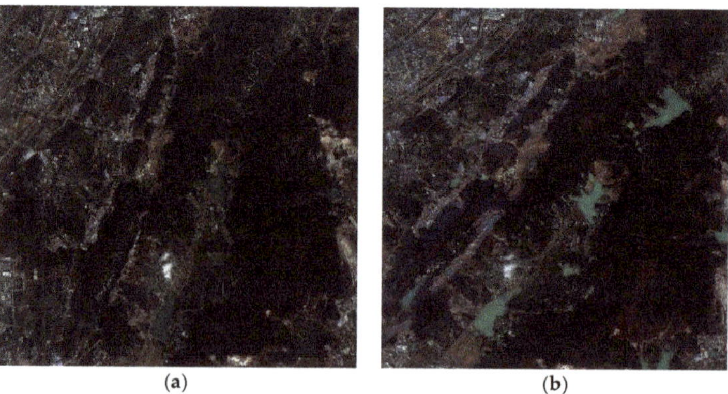

(a)   (b)

Figure 4. Comparison of multispectral images before and after fusion. (**a**) Original multispectral image; (**b**) image after fusion.

4. Band combination selection

GF-2 multispectral images contain redundant data because of the close correlation between different bands. In order to make full use of the features of GF-2 multispectral images, reduce data redundancy and maintain the original characteristics of the images, we need to make the optimal band combination for GF-2 multispectral images.

There are three principles to choose the optimal band combination: the information in a single band should be as much as possible; the information intersection between two bands should be less; the spectral differences of different types of ground objects after the band combination should be getting clearer [18]. Because the spectral bands of GF-2 multispectral images are the same as the GF-1 multispectral images, according to the above three principles, it is appropriate to use the standard deviation and Optimum Index Factor (OIF) [36] to study the best band combination of GF-2 images. Finally, we get band 2, band 3, and band 4 as the combined bands to generate the original image in our study. The remote sensing image after band combination is shown in Figure 5.

Figure 5. Image after band combination.

3.1.2. Data Labeling and Augmentation

1. Data labeling

The data labeling mainly includes two parts: one is to label a region in which the waterbody is located to get the training dataset for our proposed object detection network model and the other is to label the waterbody to get the training dataset for our FCN without pooling layers.

(a) Labeling a region: We use the LabelImg (Barcelona, Spain) to label the region in which the waterbody is located. The water area is labeled by a rectangular frame, and a xml file is generated finally. As shown in Figure 6, the label in the file records the name, path, water area category and coordinates of the frame.

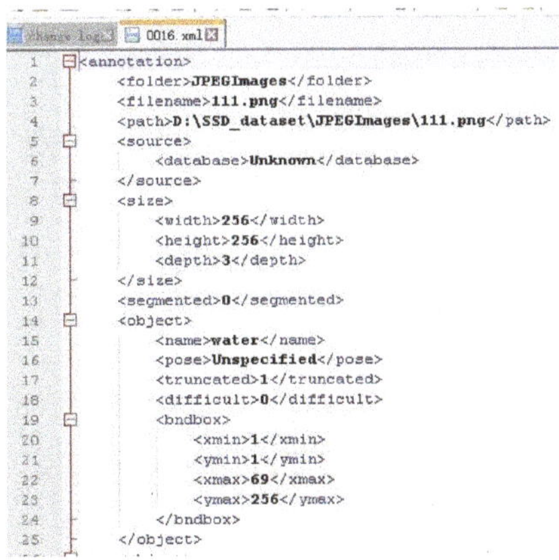

```
change log×   0016.xml×
1    <annotation>
2        <folder>JPEGImages</folder>
3        <filename>111.png</filename>
4        <path>D:\SSD_dataset\JPEGImages\111.png</path>
5        <source>
6            <database>Unknown</database>
7        </source>
8        <size>
9            <width>256</width>
10           <height>256</height>
11           <depth>3</depth>
12       </size>
13       <segmented>0</segmented>
14       <object>
15           <name>water</name>
16           <pose>Unspecified</pose>
17           <truncated>1</truncated>
18           <difficult>0</difficult>
19           <bndbox>
20               <xmin>1</xmin>
21               <ymin>1</ymin>
22               <xmax>69</xmax>
23               <ymax>256</ymax>
24           </bndbox>
25       </object>
```

Figure 6. Example of labeling a region.

(b)   Waterbody labeling: The Labelme is used to label the waterbody. The labeled image is shown in Figure 7. In the labeled image, the labeled information of the waterbody is saved in the index dataset. Because the extraction of waterbody is essentially binary classification, the black area in the labeled image is the background area and represented by 0. The red area is the waterbody and is represented by 1.

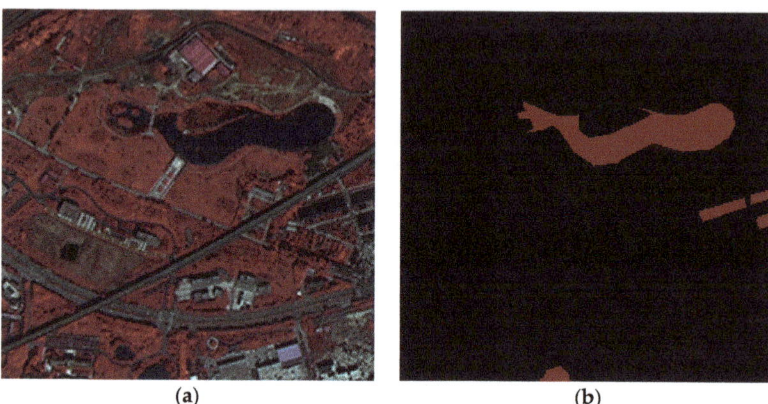

(a)   **(b)**

Figure 7. Remote sensing image and regional waterbody labeled image. (**a**) Remote sensing image; (**b**) regional waterbody labeled image.

2.   Data augmentation

Compared with public images dataset like ImageNet, there is no remote sensing image training dataset, and it is difficult to get much more data by ourselves, so we enlarge the dataset by data augmentation [37,38] to expand the training data and avoid the overfitting phenomenon. The remote sensing images are clipped, and the size of the images is 256×256, which is feasible to complete the construction of the training dataset. The geometric transformation operations used in this paper include rotation operations of $90°$, $180°$, and $270°$ of the original images, horizontal flip operation, and vertical flip operation. We use the OpenCV (Intel, Santa Clara, CA, USA) based on python for data augmentation. The operation examples are shown in Figure 8.

We labeled the data at first, and then achieve the data augmentation operations. To meet the training requirement of FCN without pooling layers, we clip the size of waterbody labeled images into 32×32. Now we have 6000 waterbody labeling data with size of 32×32, and 6000 waterbody region labeled data with size of 256×256. We choose 70% of them as the training set to train the improved water detection network and the FCN without pooling layers, respectively, and 30% are used as test data.

3.2. Improved Water Detection Network Based on YOLOv3

As shown in Figure 9, the improved water detection network based on YOLOv3 mainly includes two parts. The first part is the feature extraction module, in which we use 13 convolutional layers to obtain the water features. The second part is the feature pyramid network structure for multi-scale waterbody detection, which uses feature fusion for multi-scale waterbody detection.

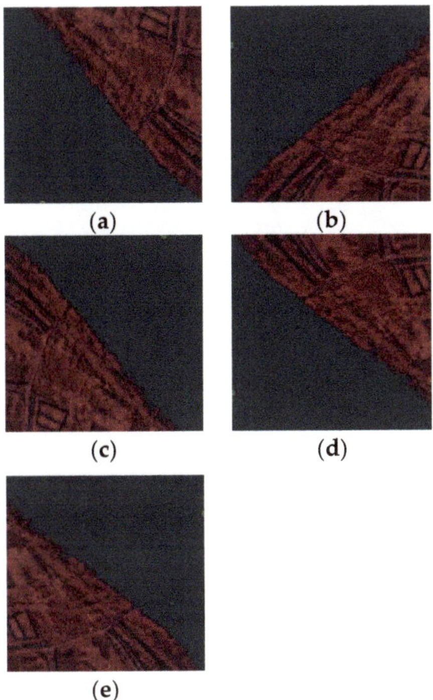

(a)

(b)

(c)

(d)

(e)

Figure 8. Example of remote sensing images by data augmentation. (**a**) 90° rotation; (**b**) 180° rotation; (**c**) 270° rotation; (**d**) horizontal flip; (**e**) vertical flip.

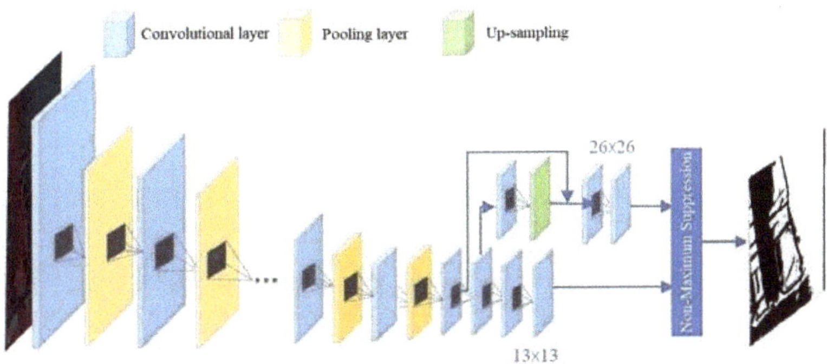

Figure 9. The network structure of the improved water detection network.

3.2.1. Improved Feature Extraction Module

The Darknet-53 network structure used in the feature extraction module of YOLOv3 (Seattle, WA, USA) easily leads to the overfitting phenomenon in the case of limited training data. In order to solve this problem, a 13-layer CNN is constructed for water feature extraction in the feature extraction module. The module is mainly composed of convolutional layers, pooling layers, and batch standardization layers. The parameters are shown in Table 1.

Table 1. Network structure and parameters of improved feature extraction module.

Type/Parameter	Number of Convolution Kernels	Convolution Kernel Size	Step	Padding
Conv1	16	1×1	1	same
Conv2	16	3×3	1	same
Conv3	256	1×1	1	same
BN				
Max Pooling				
Conv4	128	3×3	1	same
Conv5	128	1×1	1	same
Conv6	512	3×3	1	same
BN				
Max Pooling				
Conv7	256	3×3	1	same
Conv8	256	1×1	1	same
Conv9	512	3×3	1	same
BN				
Max Pooling				
Conv10	256	3×3	1	same
Conv11	256	1×1	1	same
Conv12	512	3×3	1	same
Max Pooling				
Conv13	1024	3×3	1	same

In the improved feature extraction module, the convolutional layers with a convolution kernel of 3×3 is used to extract the water features of a tidal flat area, and the convolutional layers with a convolution kernel of 1×1 is used to realize cross-channel information fusion. In order to ensure the generalization ability of our waterbody detection model in a tidal flat area, this paper uses pooling layers to keep the main characteristic data of water. To solve the slow convergence and gradient explosion, the improved feature extraction module used in this paper adds batch standardization layers. This operation normalizes the data before it passes through the activation function to reduce the change data amplitude and make it follow the Gaussian distribution and speed up the convergence of the network model as a result.

3.2.2. Feature Pyramid Network Structure for Multi-Scale Water Detection

Inspired by the design of the feature pyramid, three branches are used in YOLOv3 (Seattle, WA, USA) to obtain feature maps with sizes of 13×13, 26×26 and 52×52 respectively. The feature maps of different sizes correspond to different receptive fields. The larger the size of the feature map is, the smaller the corresponding receptive field is. The correspondence between feature graph size and prior box is shown in Table 2. Based on the size and characteristic of the tidal flat water, we design two branches in our model. One of the branches, used for the detection of narrow and long waterbodies in the tidal flat area, is to get a 13×13 feature map through three convolutional layers after the improved feature extraction module; the other branch, used for the detection of small waterbodies in tidal flat areas, is to up-sample the output of the 14th convolutional layer in the network, and then fuse it with the features obtained by the 13th convolutional layer, and finally get the feature map with size of 26×26 through two convolutional layers.

Table 2. Corresponding between feature graph size and prior box.

Size of Feature Map	Receptive Field		Prior Box		
13×13	large	116×90	156×198	373×326
26×26	middle	20×61	62×45	59×119
52×52	small	10×13	16×30	33×23

3.3. FCN without Pooling Layers

The improved object detection network based on YOLOv3 (Seattle, WA, USA) is a water object detection model, so it cannot extract the water edge. In order to solve this problem, we design the FCN without pooling layers to complete the initial waterbody extraction and obtain the feature information of waterbody in a tidal flat area. The network structure of the FCN without pooling layers is shown in Figure 10.

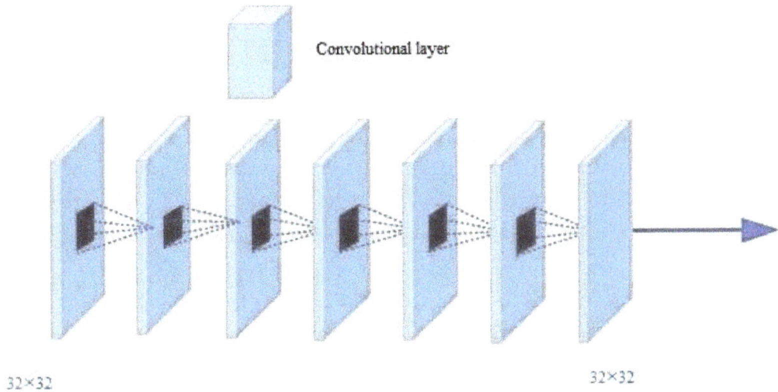

Figure 10. Fully convolutional network without pooling layers.

In general, the pooling layers of CNN have two main functions: one is to compress the extracted features to reduce the computational time of the model. The second is to enlarge the receptive field of the model so that each point in the feature map corresponds to a larger area in the original image. Because the receptive field represents the receptive range of different neurons in the network to image, the enlargement of receptive field means the enlargement of receptive range of different neurons in the network. So, each point in the feature map corresponds to a larger area in the original image when we enlarge the receptive field of the model. The FCN without pooling layers we proposed in this paper aims at the initial extraction of waterbodies from 32×32 remote sensing images, so the receptive field is not required and the water extractions based on the network still work in our method.

The FCN without pooling layers uses six convolutional layers to extract waterbodies from 32×32 remote sensing images, and the convolution kernel sizes of convolutional layers are 3×3 and 1×1, respectively. All the parameters in the network can be seen in Table 3. Compared with the convolution kernels with sizes 7×7 and 5×5, we use the convolution kernel with size 3×3 in the model to improve the network depth and the nonlinear expression ability of the model with the same receptive field. The convolution kernel with size 1×1 realizes cross-channel information fusion.

Table 3. Parameters of FCN without pooling layers.

Layer	Number of Convolution Kernels	Convolution Kernel Size	Step	Padding
Conv1	64	3×3	1	same
Conv2	64	1×1	1	same
Conv3	256	3×3	1	same
Conv4	256	1×1	1	same
Conv5	512	3×3	1	same
Conv6	512	3×3	1	same
Conv7	2	3×3	1	same

3.4. Similarity Algorithm for Water Extraction

To reduce the false extraction caused by the high similarity between the waterbody and background in a tidal flat area, a similarity algorithm for water extraction is proposed. The steps of the algorithm are as follows:

1. Firstly, we obtain the detection results of the improved water detection network based on YOLOv3 and the initial water extraction results of the FCN without pooling layers.
2. Secondly, we compute the average pixel value of the initial water extraction information obtained from the FCN without pooling layers and take it as the standard water pixel value in the tidal flat area. The formula is:

$$r, g, b = \sum_{i}^{n} L_{r,g,b}/n \tag{1}$$

where r, g, b represents the average water pixel value, $L_{r,g,b}$ represents the pixel value of the water extraction results, and n is the number of waterbody pixels.

3. Thirdly, we traverse every pixel in the detection information, and calculate the similarity between the water pixels and the standard water pixels. The formula is:

$$Y = \sqrt{(L_r - r)^2 + (L_g - g)^2 + (L_b - b)^2} \tag{2}$$

where L_r, L_g and L_b represent the pixel values of the detection results in the red, green and blue channels, respectively.

4. Finally, we set a similarity threshold and finish the water extraction. We set 34 based on the experiments. If the similarity between a water pixel in the water detection results and a standard water pixel is greater than the threshold, the pixel point is considered as water, otherwise it is not water.

The similarity algorithm for water extraction proposed in this paper effectively solves the accuracy problem of waterbody extraction caused by the blurry boundary between the waterbody and background.

The similarity algorithm for water extraction is as Algorithm 1:

Algorithm 1. Similarity Algorithm for Water Extraction.

Input: Output results of improved water detection network based on YOLOv3 and FCN without pooling layers
Output: Pixel is waterbody or non-waterbody
1. **Procedure** Similarity-Water-Extraction (n: integer);
2. **begin**
3. **for** i: = 1 to n **do**
4. **begin**
5. $sumL_r = sumL_r + L_{ri}$;
6. $sumL_g = sumL_g + L_{ri}$;
7. $sumL_b = sumL_b + L_{bi}$;
8. **end**;
9. r: = $sumL_r/n$;
10. g: = $sumL_g/n$;
11. b: = $sumL_b/n$;
12. **while** (pixel is the result of waterbody target detection) **do**
13. **begin**
14. Y = sqrt $((L_r - r)^2 + (L_g - g)^2 + (L_b - b)^2)$;
15. **if** $(Y > 34)$ **then**
16. pixel is waterbody;
17. **else then**
18. pixel is non-waterbody;
19. **end**;
20. **end**;

---

4. Experiment and Analysis

4.1. Experimental Configuration

All experiments are implemented on a system with NVIDIA GeForce GTX1070 (Santa Clara, CA, USA) and Intel(R) Core (TM) i7 (Santa Clara, CA, USA), and the operating system is Windows 10 (Redmond, WA, USA). The software environment of the system is ENVI 5.3 (Boulder, CO, USA), Python 3.6 (Wilmington, DE, USA), TensorFlow 1.12.0 (Mountain View, CA, USA) and Keras 2.2.4 (Cobham, UK).

4.2. Evaluation Criterion

To accurately analyze the experiments, this paper selects three indicators to quantitatively evaluate the model: Intersection over Union (*IoU*), pixel accuracy, and *Kappa* coefficient. The overlap ratio describes the overlap degree between the extracted object and the ground truth; the pixel accuracy is used to measure the proportion coefficient of the correct part of the detection result; the *Kappa* coefficient is used to measure the pixel classification accuracy. The calculation formulas of the three indicators are as follows:

$$IoU = \frac{Area(P) \cap Area(T)}{Area(P) \cup Area(T)} \tag{3}$$

where $Area(P)$ represents the prediction result and $Area(T)$ represents the ground truth.

$$P = \frac{TP}{TP + FP} \tag{4}$$

where P represents the pixel accuracy, TP represents the number of samples that are positive and identified as positive by the network model, and FP represents the number of samples that are incorrectly classified as positive.

$$k = \frac{p_0 - p_e}{1 - p_e} \tag{5}$$

where k represents the value of *Kappa* coefficient, p_0 represents the proportion of the correct cells, and p_e represents the proportion of misinterpretations caused by chance.

$$pe = TP * \frac{FN}{n * n} \tag{6}$$

where TP represents the number of samples that are positive and identified as positive by the network model, n represents the number of ground feature types, and FN represents the number of samples that are incorrectly classified as negative.

4.3. Parameter Setting

In this paper, the waterbody detection network plays an important role for the final water extraction. To study the influence of learning rate parameters on the accuracy of water detection, we compare and analyze the decline curve of the loss function under different learning rates and take the optimal learning rate as the model parameter at last. The values of learning rate are set as 0.0001, 0.005, 0.001 and 0.01 respectively. The convergence curve of the loss function is shown in Figure 11.

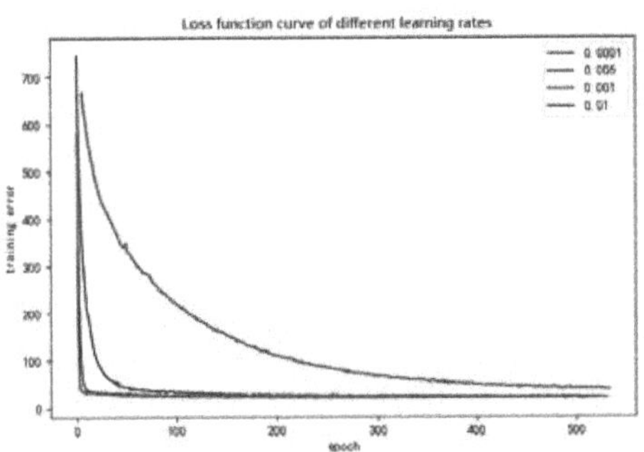

Figure 11. Loss function curve of the water detection model with different learning rates.

As shown in Figure 11, when the learning rates are 0.0001 and 0.005, the network model converges slowly, and the loss function value of the final convergence result is higher. When the learning rates are 0.001 and 0.01, the network performs better, and its convergence speed and final convergence result are significantly improved compared with other learning rates. Based on the above analysis of learning rate, as well as many experiments and model debugging, the training parameters of the water detection model are obtained. In this paper, we set the learning rate to be 0.001, the batch training sample size 64, the impulse 0.9, the weight attenuation 0.0005, and the epoch 500 for the improved water detection network based on YOLOv3 (Seattle, WA, USA). The network also uses two *IoU* thresholds during training. If a prediction overlaps the ground truth by 0.7 it is as a positive example, by 0.5–0.7 it is ignored, less than 0.5 for all ground truth objects it is a negative example. We set the learning rate to be 0.01, the batch training sample size 32, the impulse 0.9, the weight attenuation 0.0001 and the epoch 150 for the FCN without pooling layers in our experiments.

4.4. Performance Analysis

4.4.1. Influence of Threshold of Similarity Algorithm for Water Extraction

We set 31, 32, 33, 34, 35, 36, 37, 38 and 39 as thresholds, respectively, and use the extraction accuracy to study the influence of thresholds. The experiments are shown in Figure 12.

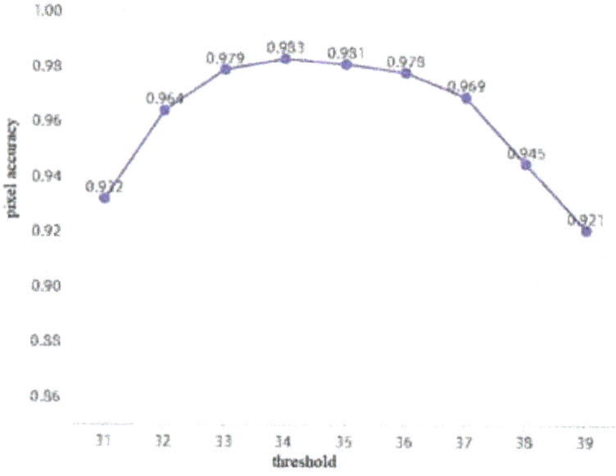

Figure 12. Comparison of the pixel accuracy with different thresholds.

We calculated the pixel accuracy of water extraction with different thresholds, and the experiments show that the pixel accuracies increasing at first and decreasing after 34, as shown in Figure 12. When the threshold is getting 34, the pixel accuracy is the highest in our experiments. When the threshold is lower than 34, the phenomenon of missing extraction begins to appear in the water extraction, which makes the accuracy of the water extraction continue to decrease. When the threshold is higher than 34, water extraction begins to appear the false extraction, and the accuracy of water extraction decreases with the increase of the threshold as well. This is likely caused by the definition of the standard water pixel value. To sum up, this paper selects 34 as the threshold of similarity algorithm for water extraction in the tidal flat area.

4.4.2. Qualitative Analysis

To verify the effectiveness of this method, we compare the following methods: NDWI, support vector machine (SVM), maximum likelihood classification, U-Net (Freiburg, Germany) [39], YOLOv3 (Seattle, WA, USA) and FYOLOv3. The tidal flat remote sensing images from the GF-2 satellite are selected as the sample, and the experiments are shown in Figure 13.

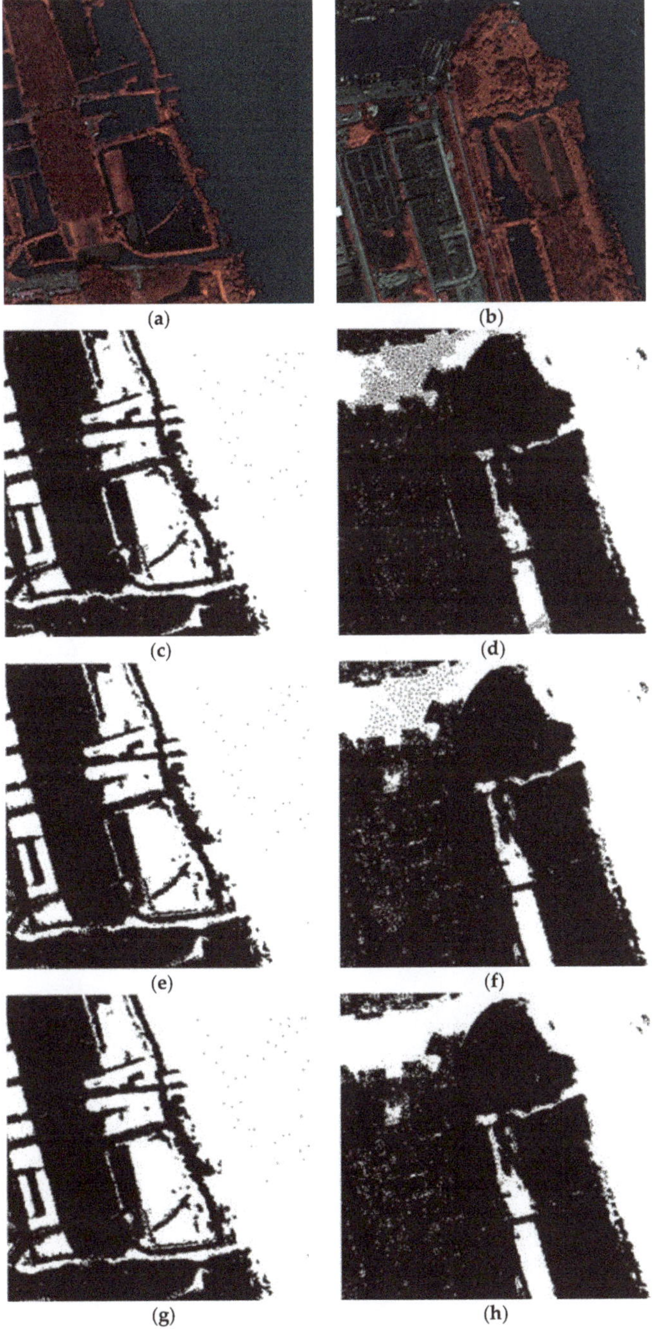

Figure 13. *Cont.*

Figure 13. Comparison of different methods for water extraction in a tidal flat area. (**a**) Small water bodies; (**b**) water bodies with blurry boundaries; (**c,d**) NDWI; (**e,f**) SVM; (**g,h**) maximum likelihood classification; (**i,j**) U-Net; (**k,l**) YOLOv3; (**m,n**) FYOLOv3.

The experiments of NDWI, SVM, maximum likelihood classification, U-Net (Freiburg, Germany), YOLOv3 (Seattle, WA, USA) and FYOLOv3 for small waterbodies and waterbodies with blurry boundaries are shown in Figure 13, respectively.

As shown in Figure 13, the NDWI method effectively extracts the waterbody in the remote sensing images, but there are a lot of noises in the extraction results. The water extractions by SVM and maximum likelihood classification method are relatively good, but they cannot effectively solve the problem of high similarity between water and background, and there are a lot of false extractions in the experiments. From the experiments of U-Net, we can see that the water extraction is not good for small waterbodies, and there are lots of false extraction and missing extraction. Compared to NDWI, SVM, maximum likelihood classification, and U-Net (Freiburg, Germany), the experiments of YOLOv3 (Seattle, WA, USA) and FYOLOv3 have better extraction. However, in the experiments of YOLOv3 (Seattle, WA, USA), there are some missing extractions in the densely small water areas due to the prior frames, and FYOLOv3 is able to check each pixel in the detection area based on the similarity algorithm for water extraction, which solves the false and missing extraction, so it is superior to YOLOv3 (Seattle, WA, USA).

4.4.3. Accuracy Analysis

We take *IoU*, pixel accuracy (*P*) and *Kappa* (*k*) coefficient as the evaluation indexes to compare six methods: NDWI, SVM, maximum likelihood classification, U-Net (Freiburg, Germany), YOLOv3 (Seattle, WA, USA), and FYOLOv3. We set the image size to be 256×256, the learning rate 0.001, the decay 0.0005, the momentum 0.9 and we use the optimizer Adam for YOLOv3. We set the learning rate to be 0.001, the decay 0.0001 the momentum 0.9 and the optimizer Adam for U-Net (Freiburg, Germany). The threshold of NDWI is 0.19 and the parameter of maximum likelihood is 2.1. The experiment results of six methods are shown in Table 4.

Table 4. Accuracy comparison of six methods for water extraction in tidal flat area.

Method	*IoU*	*P*	*k*
NDWI	0.9351	0.9665	0.9303
SVM	0.8925	0.9432	0.8821
Maximum likelihood classification	0.9041	0.9496	0.8952
U-Net	0.9309	0.9642	0.9251
YOLOv3	0.9436	0.9710	0.9394
FYOLOv3	0.9679	0.983	0.9613

As shown in Table 4, *IoU*, *P* and *k* of the FYOLOv3 method for water extraction in a tidal flat area on remote sensing images are the highest, followed by the YOLOv3 network, NDWI, U-Net, maximum likelihood classification, and SVM. The method proposed in this paper has higher extraction accuracy than other methods and has a better effect for water extraction in tidal flat with fuzzy boundaries and small waterbodies in a tidal flat area. This proves that this method has more advantages for small waterbody extraction in a tidal flat area.

Table 5 shows the model training time and water extraction time of the three convolutional neural network methods. Although the FYOLOv3 method is divided into three parts, its speed of water extraction is the highest. The method proposed in this paper not only improves the accuracy of water extraction, but also reduces the model training time and water extraction time due to the improvement of YOLOv3 (Seattle, WA, USA).

Table 5. Comparison of the model training time and water extraction time.

Method	Training Time (h)	Water Extraction Time (s/Sheet)
U-Net	8	0.18
YOLOv3 + FCN + Similarity algorithm	$(10 + 3)$	$0.24 \times (0.11 + 0.08 + 0.05)$
FYOLOv3	$(6 + 3)$	$0.16 \times (0.03 + 0.08 + 0.05)$

5. Conclusions

The tidal flat is long and narrow with high sediment content, so there is little feature difference between the waterbody and the background, and the boundary of the waterbody is blurry. Extracting tidal flat waterbody accurately from high-resolution remote sensing imagery is a great challenge. In order to solve the low accuracy problem of tidal flat waterbody extraction, in this paper, a FYOLOv3 is proposed to solve the above problems and extract waterbody in tidal flat with high accuracy. The FYOLOv3 mainly includes three parts: Firstly, according to the characteristics of tidal flat water extraction, an improved object detection network based on YOLOv3 (Seattle, WA, USA) is proposed to ensure the accuracy of water detection, reduce the computational time of the model and alleviate the overfitting phenomenon; secondly, a FCN without pooling layers follows the improved object detection network to obtain the initial water extraction; at last, a similarity algorithm for water extraction is proposed, which distinguishes the waterbody and non-water pixel by pixel in order to improve the extraction accuracy of tidal flat waterbody. Compared to the other models, the experiments show that our method has higher accuracy on the waterbody extraction of tidal flats or small areas, and the *IoU* of our method is 2.43% higher than YOLOv3 (Seattle, WA, USA) and 3.7% higher than U-Net (Freiburg, Germany). However, this method also has some limitations, which needs to manually select the similarity threshold, and different thresholds need to be set for different data, which affects the robustness of the method. Therefore, our future research will consider how to determine the threshold intelligently in order to improve the robustness of the method.

Author Contributions: Conceptualization, L.Z. and Y.F.; Methodology, L.Z. and Y.F.; Validation, Y.S. and R.Y.; Resources, R.Y. and G.W.; Data Curation, J.W. and G.W.; Writing—Original Draft Preparation, L.Z. and Y.F.; Writing—Review and Editing, L.Z. and Y.F.; Supervision, L.Z.; Funding Acquisition, L.Z. and G.W. All authors have read and agreed to the published version of the manuscript.

Funding: This research was funded by the National Key Research and Development Program of China (No.2016YFA0601703, 2016YFC0401005) and National Natural Science Foundation of China (91847301, 42075191, 52009080).

Institutional Review Board Statement: Not applicable.

Informed Consent Statement: Not applicable.

Data Availability Statement: The study did not report any data.

Conflicts of Interest: The authors declare no conflict of interest.

References

1. Shao, Y.; Zhou, J.; Mu, R.; Zhu, L.; Jiang, T.Y. Research on urban development and wetland protection in China. *J. Ecol. Environ.* **2018**, *27*, 381–388.
2. Li, L.; Zhao, J.; Xue, X.F. Comprehensive treatment and water quality simulation of Nanming River in Guizhou. *J. Environ. Sci.* **2018**, *38*, 1920–1928.
3. Jin, J.; Li, G.; Sun, W.; Yang, X.; Chang, X.; Liu, K.; Liu, Y. Application status and Prospect of satellite remote sensing water resources survey and monitoring. *Surv. Mapp. Bull.* **2020**, *0*, 7–10.
4. Rasid, H.; Pramanik, M.A.H. Visual interpretation of satellite imagery for monitoring floods in Bangladesh. *Environ. Manag.* **1990**, *14*, 815–821. [CrossRef]
5. Feyisa, G.L.; Meilby, H.; Fensholt, R.; Proud, S.R. Automated water extraction index: A new technique for surface water mapping using landsat imagery. *Remote Sens. Environ.* **2014**, *140*, 23–35. [CrossRef]
6. Yao, F.; Wang, C.; Dong, D.; Luo, J.; Shen, Z.; Yang, K. High-Resolution mapping of urban surface water using ZY-3 multi-spectral imagery. *Remote Sens.* **2015**, *7*, 12336–12355. [CrossRef]
7. Guo, Q.; Pu, R.; Li, J.; Cheng, J. A weighted normalized difference water index for water extraction using Landsat imagery. *Int. J. Remote Sens.* **2017**, *38*, 5430–5445. [CrossRef]
8. Xu, H. Extraction of water information by improved normalized difference water index (MNDWI). *Chin. J. Remote Sens.* **2005**, *5*, 589–595.
9. McFeeters, S.K. The use of the Normalized Difference Water Index (NDWI) in the delineation of open water features. *Int. J. Remote Sens.* **1996**, *17*, 1425–1432. [CrossRef]
10. Zhang, J.; He, C.; Pan, Y.; Li, J. Classification of high spatial resolution remote sensing data based on multi-source information combination of SVM. *J. Remote Sens.* **2006**, *10*, 49–57.
11. Sisodia, P.S.; Tiwari, V.; Kumar, A. Analysis of supervised maximum likelihood classification for remote sensing images. In Proceedings of the IEEE International Conference on Recent Advances and Innovations in Engineering (ICRAIE-2014), Jaipur, India, 9–11 May 2014; pp. 1–4.
12. Duan, Q.; Meng, L.; Fan, Z.; Hu, W.; Xie, W. Applicability of water information extraction method from GF-1 satellite images. *Land Resour. Remote Sens.* **2015**, *27*, 79–84.
13. Kirby, R. Practical implications of tidal flat shape. *Cont. Shelf Res.* **2000**, *20*, 1061–1077. [CrossRef]
14. Jain, S.K.; Singh, R.D.; Jain, M.K.; Lohani, A.K. Delineation of Flood-Prone Areas Using Remote Sensing Techniques. *Water Resour. Manag.* **2005**, *19*, 333–347. [CrossRef]
15. Wang, J.; Zhang, Y.; Kong, G. The application of multi-band spectral relationship method in waterbody extraction. *Mine Surv.* **2004**, *4*, 30–32.
16. Hinton, G.E.; Salakhutdinov, R.R. Reducing the dimensionality of data with neural networks. *Science* **2006**, *313*, 504–507. [CrossRef] [PubMed]

17. Krizhevsky, A.; Sutskever, I.; Hinton, G.E. Imagesnet classification with deep convolutional neural networks. *Adv. Neural Inf. Process. Syst.* **2012**, *25*, 1097–1105.
18. Zhong, X. Research on Waterbody Recognition from Remote Sensing Images Based on Convolution Neural Network. Master's Thesis, Hohai University, Nanjing, China, 2019.
19. Liang, Z. Extraction Method of Multi-Source Remote Sensing Water Information Based on Deep Learning and Its Application. Master's Thesis, Anhui University, Hefei, China, 2019.
20. Song, S.; Liu, J.; Liu, Y.; Feng, G.; Han, H.; Yao, Y.; Du, M. Intelligent object recognition of urban water bodies based on deep learning for multi-source and multi-temporal high spatial resolution remote sensing imagery. *Sensors* **2020**, *20*, 397. [CrossRef] [PubMed]
21. Yu, L.; Wang, Z.; Tian, S.; Ye, F.; Ding, J.; Kong, J. Convolutional neural networks for waterbody extraction from Landsat imagery. *Int. J. Comput. Intell. Appl.* **2017**, *16*, 1750001. [CrossRef]
22. Li, L.; Yan, Z.; Shen, Q.; Cheng, G.; Gao, L.; Zhang, B. Waterbody extraction from very high spatial resolution remote sensing data based on fully convolutional networks. *Remote Sens.* **2019**, *11*, 1162. [CrossRef]
23. Wang, Z.; Gao, X.; Zhang, Y.; Zhao, G. MSLWENet: A novel deep learning network for lake water body extraction of Google remote sensing images. *Remote Sens.* **2020**, *12*, 4140. [CrossRef]
24. Yu, Y.; Yao, Y.; Guan, H.; Li, D.; Liu, Z.; Wang, L.; Yu, C.; Xiao, S.; Wang, W.; Chang, L. A self-attention capsule feature pyramid network for water body extraction from remote sensing imagery. *Int. J. Remote Sens.* **2021**, *42*, 1801–1822. [CrossRef]
25. Li, L.; Wen, Q.; Wang, B.; Fan, S.; Li, L.; Liu, Q. Water body extraction from high-resolution remote sensing images based on scaling efficientNets. *J. Phys. Conf. Ser.* **2021**, *1894*, 012100. [CrossRef]
26. Redmon, J.; Divvala, S.; Girshick, R.; Farhadi, A. You only look once: Unified, real-time object detection. In Proceedings of the IEEE Conference on Computer Vision and Pattern Recognition, Las Vegas, NV, USA, 27–30 June 2016; pp. 779–788.
27. Redmon, J.; Farhadi, A. YOLO9000: Better, faster, stronger. In Proceedings of the IEEE Conference on Computer Vision and Pattern Recognition, Honolulu, HI, USA, 21–26 July 2017; pp. 7263–7271.
28. Redmon, J.; Farhadi, A. YOLOv3: An incremental improvement. *arXiv* **2018**, arXiv:1804.02767.
29. Szegedy, C.; Liu, W.; Jia, Y.; Sermanet, P.; Reed, S.; Anguelov, D.; Erhan, D.; Vanhoucke, V.; Rabinovich, A. Going deeper with convolutions. In Proceedings of the IEEE Conference on Computer Vision and Pattern Recognition, Boston, MA, USA, 7–12 June 2015; pp. 1–9.
30. Montanaro, M.; Lunsford, A.; Tesfaye, Z.; Wenny, B.; Reuter, D. Radiometric calibration methodology of the Landsat 8 thermal infrared sensor. *Remote Sens.* **2014**, *6*, 8803–8821. [CrossRef]
31. Huang, X.; Zhu, J.; Han, B.; Jamet, C.; Tian, Z.; Zhao, Y.; Li, J.; Li, T. Evaluation of four atmospheric correction algorithms for GOCI images over the Yellow Sea. *Remote Sens.* **2019**, *11*, 1631. [CrossRef]
32. Liu, S.; Wan, J.; Zhang, J.; Ma, Y.; Ren, G. Orthorectification method of SPOT5 satellite image based on ERDAS. *Oceanography* **2008**, *28*, 30–33.
33. Li, H.; He, X.; Tao, D.; Tang, Y.; Wang, R. Joint medical image fusion, denoising and enhancement via discriminative low-rank sparse dictionaries learning. *Pattern Recognit.* **2018**, *79*, 130–146. [CrossRef]
34. Fan, W.; Li, H.; Wen, Q.; Gao, X. Orthorectification accuracy analysis of GF-2 satellite image. *Surv. Mapp. Bull.* **2016**, *0*, 63–66.
35. Sun, W.; Chen, B.; Messinger, D. Nearest-neighbor diffusion-based pan-sharpening algorithm for spectral images. *Opt. Eng.* **2014**, *53*, 013107. [CrossRef]
36. Ren, J.; Yang, W.; Deng, X.; Wang, L.; Wang, F. Applicability of gf-2 image classification based on OIF and optimal scale segmentation. *Mod. Electron. Technol.* **2018**, *41*, 72–77, 82.
37. Shorten, C.; Khoshgoftaar, T.M. A survey on images data augmentation for deep learning. *J. Big Data* **2019**, *6*, 60. [CrossRef]
38. Ian, G.; Yoshua, B.; Aaron, C. *Deep Learning*; People's Posts and Telecommunications Press: Beijing, China, 2017; pp. 208–209.
39. Ronneberger, O.; Fischer, P.; Brox, T. U-net: Convolutional networks for biomedical images segmentation. In Proceedings of the International Conference on Medical Images Computing and Computer-Assisted Intervention, Munich, Germany, 5–9 October 2015; Springer: Cham, Switzerland, 2015; pp. 234–241.

Article

A Wide Area Multiview Static Crowd Estimation System Using UAV and 3D Training Simulator

Shivang Shukla *,†, Bernard Tiddeman † and Helen C. Miles †

Department of Computer Science, Aberystwyth University, Aberystwyth SY23 3DB, UK; bpt@aber.ac.uk (B.T.); hem23@aber.ac.uk (H.C.M.)
* Correspondence: shs36@aber.ac.uk
† S.S. wrote the software, conducted the experiments and wrote the paper. B.T. and H.C.M. advised and supervised S.S., helped with the experimental design, and proof read the paper.

Abstract: Crowd size estimation is a challenging problem, especially when the crowd is spread over a significant geographical area. It has applications in monitoring of rallies and demonstrations and in calculating the assistance requirements in humanitarian disasters. Therefore, accomplishing a crowd surveillance system for large crowds constitutes a significant issue. UAV-based techniques are an appealing choice for crowd estimation over a large region, but they present a variety of interesting challenges, such as integrating per-frame estimates through a video without counting individuals twice. Large quantities of annotated training data are required to design, train, and test such a system. In this paper, we have first reviewed several crowd estimation techniques, existing crowd simulators and data sets available for crowd analysis. Later, we have described a simulation system to provide such data, avoiding the need for tedious and error-prone manual annotation. Then, we have evaluated synthetic video from the simulator using various existing single-frame crowd estimation techniques. Our findings show that the simulated data can be used to train and test crowd estimation, thereby providing a suitable platform to develop such techniques. We also propose an automated UAV-based 3D crowd estimation system that can be used for approximately static or slow-moving crowds, such as public events, political rallies, and natural or man-made disasters. We evaluate the results by applying our new framework to a variety of scenarios with varying crowd sizes. The proposed system gives promising results using widely accepted metrics including MAE, RMSE, Precision, Recall, and F1 score to validate the results.

Keywords: crowd estimation; 3D simulation; unmanned aerial vehicle; synthetic crowd data

Citation: Shukla, S.; Tiddeman, B.; Miles, H.C. A Wide Area Multiview Static Crowd Estimation System Using UAV and 3D Training Simulator. *Remote Sens.* **2021**, *13*, 2780. https://doi.org/10.3390/rs13142780

Academic Editors: Anwaar Ulhaq and Douglas Pinto Sampaio Gomes

Received: 28 May 2021
Accepted: 9 July 2021
Published: 15 July 2021

Publisher's Note: MDPI stays neutral with regard to jurisdictional claims in published maps and institutional affiliations.

1. Introduction

Crowd estimation refers to the practice of calculating the total number of people present in a crowd. Manual crowd estimation and automated crowd estimation are the two most common broad approaches to measuring crowd size, but the method varies according to the crowd size. Manually monitoring and estimating a small crowd by splitting people into groups is a traditional way that still exists. However, manual estimation of a large crowd is not possible and may be very expensive and time-consuming. It has prompted scientists and researchers from various disciplines across the globe to develop automated crowd estimation systems that calculate the number of people in a large crowd. In the last five years, the domain has expanded rapidly. The introduction of deep learning methods, coupled with easy availability of powerful GPU based systems, has provided a step change in computer vision algorithms across a range of problem domains, starting with classification, but it has quickly moved on to other areas such as crowd estimation. A number of well-publicized crowd-related incidents and gatherings have drawn the attention of researchers and the computer vision community; it has prompted them to develop accurate crowd surveillance systems.

An example application is for use in major disasters. In such a scenario, a crowd estimation system would give a more accurate picture of the crowd and number of affected people and their geographical spread. This would enable proper coordination of the disaster teams, leading to more efficient relief-aid work.

Crowd estimation systems using Unmanned Aerial Vehicles (UAVs) is an emerging research area, due to its potential to cover a wide area in a short period. However, it presents automated estimation issues. Both the camera and the crowd are likely to be moving, so there is a risk of multiple counting of the same person. Most of the existing automated methods focus on individual frames from a single static camera. Recently, there has been some promising research conducted in multiple view systems and UAV-based cameras. UAVs can cover a large area. However, they pose problems of (1) a moving camera, (2) the crowd that may move during the capture time and (3) different view points which require extensive additional training and testing data.

Given the challenges of gathering and annotating data, our paper explores the use of a simulator to generate the training images and annotated ground-truth data. Furthermore, we have introduced a novel automated 3D crowd estimation system using a UAV, that was trained and tested with our simulator. In the initial phase, we have focused on the problem of static crowds and intend to move towards more dynamic crowds in the future. The motivation for developing our system is for crowd flow management, large-scale public gathering monitoring, public event security and relief-aid work by welfare organisations in disaster-hit areas.

This paper covers the following contributions:

1. We have extensively studied the existing crowd estimation methods, data sets, and open-source crowd simulators, along with an assessment of their shortcomings. We have focused on the intended use to identify the need to develop a new simulator for estimating the crowd using UAV.
2. We have explored in detail the development of a new 3D crowd simulation system that can generate the required training images and annotated ground truth data. Furthermore, we have generated various 3D models along with accompanying camera locations and orientations.
3. We have trained, tested and validated the simulation system against real-crowd data, where we have tested synthetic data against real crowd data sets using various state-of-the-art methods. Furthermore, we have trained a new model based on our aerial synthetic data and tested it against the real-crowd data.
4. We have introduced a novel 3D crowd estimation technique using UAV for a robust and accurate estimation of a crowd spread over a large geographical area. Our proposed solution overcomes the issue of counting of the same individual multiple times from a moving UAV.
5. We also discuss the remaining challenges for wide area crowd estimation and suggest future directions for research. Additionally, we have covered significant issues for aerial crowd data collection and have put across some promising research challenges that needs to be explored.

The remainder of this paper is organized as follows:

Section 2 provides an up-to-date review of the most relevant recent literature including recently introduced crowd estimation methods. Section 3 provides a detailed step-by-step discussion of our approach and describes the benchmarks used for evaluation. In Section 4, we discuss the implementation and setup of our system. In Section 5, we present and analyze the results of our experimental evaluation of the system. The work presented in the paper is summarized and the results and their interpretation have been discussed in Section 6. Finally, we discuss the conclusion and future directions in Section 7.

2. Related Work and Scope

2.1. Manual Crowd Estimation

In 1967, Herbert Jacobs [1], a professor at the University of California, Berkeley, proposed a simple method of dealing with scenarios where estimating the crowd size is not as easy as counting the number of tickets sold. His office looked out onto the plaza where students assembled to protest the Vietnam War. The concrete on this plaza was divided into grids. Jacobs used the layout to develop his method for estimating crowd size based on area times density. As he observed numerous demonstrations, Jacobs gathered plenty of information that led him to come up with a few basic rules of thumb, which are still used today. According to Jacobs, in a loose crowd, where the distance between each person is about one arm's length, one person would occupy 10 square feet of space. People occupy 4.5 square feet for a dense crowd, and 2.5 square feet for a mosh-pit density crowd.

In other words, if you knew the area that the crowd was covering and you applied Jacob's rule of thumb for the density of the crowd, you could easily estimate the size of the crowd by multiplying the area by density. In practice, however, it is not always easy to determine the specific area crowds cover, and densities may vary across a crowd. Suppose a crowd has gathered to hear a speaker up on a stage. We might predict that the crowd would be denser up front and less dense during the back and around the edges. To address these problems, it may be helpful to divide the crowds into low, medium, and high density zones and collect samples from each. The sample method would allow us to obtain a more reliable representation of the crowd area and density along with an estimate of standard error for both. We can use the delta rule to find the relative standard error for our estimation of the crowd size, if we have the standard errors for area and density and assume they are at least roughly independent.

Manually detecting the development and movement of a crowd around the clock, or manually counting persons in exceptionally dense crowds, is a time-consuming process. When it comes to static linear and static nonlinear events, where the entire crowd is present at the same time in a single-session event, such as a Christmas Parade or a Pride Parade, there is a higher chance of getting a false estimate due to a shift in the crowd and counting the same person multiple times. Manual crowd estimating techniques like Jacob's Crowd Formula (JCF) are inefficient in dealing with such a large flow of crowd, these methods are confined to finding the average of the overlapped or shifting crowd sizes. Thus, it is very likely to estimate a larger crowd size than expected that would result in a crowd count with an unknown error rate. Considering the challenging situations such as dynamic linear or dynamic nonlinear events, it is extremely difficult to count and maintain an accurate estimate as these events often have free-flowing crowds with various entry points and can be stretched across several sessions or days. These methods are suitable for estimating the maximum crowd capacity in an area, but when it comes to accurate estimation, there is a need to develop automatic crowd estimation methods.

2.2. Computer Vision for Crowd Estimation and Analysis

Computer vision-based crowd estimation has gained considerable attention in various aspects of crowd analysis. In 2017, Marsden et al. [2] described that crowd analysis focuses on developing task-specific systems that perform a single analysis task such as crowd counting, crowd behavior recognition, crowd density level classification and crowd behavior anomaly detection. For crowd estimation purposes, crowd counting approaches may vary based on factors like estimating the crowd from an image or from a real-time video. Loy et al. [3] classified the crowd counting approaches into three different categories known as detection-based [4,5], regression-based [6,7] and density-based estimation [8]. The evolving interest of researchers in the last five years has contributed to new developments and rapid expansion in the crowd counting domain, where the researches have mainly concentrated on crowd tracking, pedestrian counting and crowd behavior analysis, among other tasks.

Idrees et al. [9] performed dense crowd estimation using a locally consistent scale to capture the similarity between local neighbourhoods and its smooth variation using images. The high crowd density and challenging nature of the data set led to several failure cases. A high-confidence detection in the first iteration often made the method over-sensitive to detection hypotheses occurring at the desired scale in neighboring areas. Similarly, at early iterations, high confidence nonhuman detection drastically degraded the prior scale, because they provided incorrect scale information. It led to misdetections in the surrounding areas that later papers tried to address.

Zhang et al. [10] proposed a multicolumn convolutional neural network architecture (MCNN) which could estimate the number of people in a single image. By creating a network comprising of three columns corresponding to filters with receptive fields of varying sizes, the proposed approach offers resistance to huge variations in object scales (large, medium, small). The three columns were created to accommodate various object scales in the images. In addition, it offered a novel method to generate ground truth crowd density maps. In contrast to the existing methods that either summed up Gaussian kernels with a fixed variance or perspective maps, Zhang et al. also proposed that perspective distortion should be taken into consideration by estimating the spread parameter of the Gaussian kernel based on the size of each person's head within the image. However, using density maps to determine head sizes and their underlying relationships is impractical. Instead, the authors employed a key feature noticed in high-density crowd images: the relationship between head size and distance between the centres of two neighbouring people. Each person's spread parameter was calculated using data-adaptive methods based on their average distance from their neighbours. It is worth noting that the ground truth density maps generated using this method included distortion information without employing perspective maps.

Zhang et al. [11] recently introduced a multiview crowd counting method using 3D features fused with 3D scene-level density maps. The deep neural network-based (DNN) 3D multiview counting method was integrated with camera views to estimate the 3D scene-level density maps. This method used 3D projection and fusion, which could address situations where people were not all at the same height (e.g., people standing on a staircase) and provided a way to tackle the scale variation issue in 3D space without a scale selection operation. However, increasing the height resolution did not contribute to the body's information, but could introduce more noise (other people's features) along the z-dimension, resulting in poor performance.

Zhao et al. [12] introduced crowd counting with limited supervision. Initially, it labeled the most informative images and later introduced a classifier to align the data and then performed estimation based on density. The number of labeled individuals varied over the course of trials and cycles. However, the ground truth was unknown, so it was difficult to determine the exact number of people, which led to a higher or lower detection rate. A ground truth verification is necessary to overcome the problem or justify the introduced method since labeling more or fewer heads does not imply a better or worse performance.

Recently, Wang et al. [13] developed a new Scale Tree Network (STNet) for crowd counting that aimed to improve scale variety and pixel-wise background recognition. The STNet network consistently met the challenges of drastic scale variations, density changes, and complex backgrounds. A tree-based scale enhancer dealt with scale variations and a multilevel auxiliator filtered pixels from complex backgrounds and adjusted them to density changes. STNet proved to be superior to the state-of-the-art approaches on four popular crowd counting data sets, while employing fewer parameters at the same time. They also proposed a method for addressing the crowd and background imbalance problem using pure background images. This concept could be easily incorporated in other crowd counting algorithms to further improve accuracy.

Ranjan et al. [14] recently published a crowd counting method based on images with imprecise estimation. The majority of the presented work focused on estimating crowd density and using a random sample selection technique to eliminate the need for labeled data. They provided results that showed improved performance based on selecting only 17% of the training samples previously used.

Mustapha et al. [15] presented a study that used CNN and Support Vector Machines (SVM) with sensor data adapted from both structure sensors and accelerometers of wearable devices to study crowd flows and bridge loads. A classification was used to determine crowd flow classification either as a binary choice of motion speed being fast or slow or as a multiclass decision based on high, medium, low, heavy, and light crowd loads, with heavy and light corresponding to crowd load designation. The load estimate of the crowd on the structure was calculated using regression to obtain the overall weight in kilograms. However, the regression results revealed inconsistency in fusion performance and a huge percentage of errors, when using the raw signal for SVM. Additionally, the study was conducted on a small scale. While considering the size of the crowd, however, any size can be considered in the future. That said, a large-scale crowd flow study is required to establish and comprehend the relationship between crowd flow and bridge load.

Almeida et al. [16] recently proposed a crowd flow filtering method to analyze crowd flow behavior. It converted the input for the optical flow from an image plane into world coordinates to perform a local motion analysis, while exploring the Social Forces Model. The filtered flow was then returned to the image plane. The method was evaluated using an image plane and needs to be expanded for the image's analysis to world coordinates. However, the work was confined to static cameras and could monitor behavior in a limited area. In addition, there is a pressing need to implement the proposed filtering approach on GPUs to achieve even faster execution times. However, the possibility for substantial speedups must be assessed.

Choi et al. [17] recently presented 3DCrowdNet, a 2D human pose-guided 3D crowd pose and shape estimation system for in-the-wild scenes. The 2D human pose estimation methods provide relatively robust outputs on crowd scenes than 3D human pose estimation methods. After all, they can exploit in-the-wild multiperson 2D data sets. Nevertheless, the challenge remains in recovering accurate 3D poses from images with close interaction. Extreme instances frequently entail difficult poses and substantial interperson occlusion, which are both uncommon in the existing trained data.

Fahad et al. [18] attempted to address the issue of public venues by using static camera positions that only record the top view of the images. To deal with events like strikes and riots, the proposed approach captured both the top and front view of the photos. The congested scene recognition (CSRNet) model assessed in this study utilized two separate test cases, one with only top view photos and the other with only front view images. However, the mean absolute error (MAE) and mean squared error (MSE) values of the front view images were higher than the top view images which needs to be reconsidered using other state-of-the-art networks. The gradient adversarial neural network (GANN) network could be effective in resolving the problem of projecting images from multiple viewpoints.

2.3. Previous Reviews and Surveys

Zhan et al. [19] presented the first assessment of crowd analysis approaches used in computer vision research and discussed how diverse research disciplines can assist computer vision approach. Later on, Junior et al. [20] provided a survey on crowd analysis using computer vision techniques, which covered topics including people monitoring, crowd density estimation, event detection, validation, and simulation. The research focused on three key issues in crowd analysis: density estimation, tracking in crowded settings, and analysing crowd behavior at a higher level, such as temporal evolution, primary directions, velocity predictions, and detection of unexpected situations. In terms of crowd synthesis, the review mostly focused on crowd models that either used computer vision algorithms to extract real-world data to improve simulation realism or were used to train and test computer vision techniques.

Teixeira et al. [21] presented the first human sensing survey offering a comprehensive analysis of the presence, count, location, and track of a crowd. It focused on five commonly encountered spatio-temporal properties: identity, presence, count, location, and track. The survey provided an inherently multidisciplinary literature of human-sensing, focusing mainly on the extraction of five commonly needed spatio-temporal properties: namely presence, count, location, track and identity. It also covered a new taxonomy of observable human attributes and physical characteristics, as well as the sensing technologies that may be utilized to extract them. In addition, it compared active and passive sensors, sensor fusion techniques, and instrumented and uninstrumented settings.

Loy et al. [3] discussed and evaluated state-of-the-art approaches for crowd counting based on video images as well as a systematic comparison of different methodologies using the same procedure. The review concluded that regression models capable of dealing with multicollinearity among features, such as Kernel ridge regression (KRR), Partial least-squares regression (PLSR), and Least-squares support vector regression (LSSVR), perform better than linear regression (LR) and random forest regression (RFR). The findings also revealed that depending on the crowd structure and density, certain features may be more useful. In sparse settings, foreground segment-based features could give all of the information required to estimate crowd density. Edge-based features and texture-based features, on the other hand, became increasingly important when a scene becomes packed with frequent interobject occlusions. Depending on the data set and regression model used, the final results affirmed that combining all attributes does not always help.

In 2014, Ferryman et al. [22] presented a PETS2009 crowd analysis data set and highlighted performance in detection and tracking. It first published a performance review of state-of-the-art crowd image analysis visual surveillance technologies, using defined metrics to objectively evaluate their detection and tracking algorithms. Comparing results with others, whether anonymous or not, was a practical and encouraging research strategy for advanced, robust, real-time visual systems. Furthermore, the latest findings highlighted the requirement for ground truth data sets, which may be used to showcase the different systems capabilities, such as accuracy, precision, and robustness.

Li et al. [23] examined the state-of-the-art techniques for crowded scene analysis in three major areas: motion pattern segmentation, crowd behavior recognition, and anomaly detection, using various methods such as crowd motion pattern learning, crowd behavior, activity analysis, and anomaly detection in crowds. The survey concluded that crowded settings frequently involve extreme clutter and object occlusions, making current visual-based techniques difficult to use. Fusion of data from several sensors is a proven tool to eliminate confusion and enhance accuracy [24]. Another finding revealed that many existing video analysis systems track, learn, and detect by integrating the functional modules, without taking into account the interactions between them. It was preferable for crowded scene analysis systems to execute tracking, model learning, and behavior recognition in a completely online and unified manner to effectively utilise the hierarchical contextual information. Despite the development of several approaches for feature extraction and model learning in crowded scene analysis, there is no widely acknowledged crowded scene representation.

Ryan et al. [25] offered a comparison of holistic, local, and histogram-based approaches as well as numerous picture characteristics and regression models, across multiple data sets. The performance of five public data sets was evaluated using a *K*-fold cross-validation protocol: the UCSD [26], PETS 2009 [27], Fudan [28], Mall [29], and Grand Central [30] data sets. The survey of the various methods concluded that the usage of local features consistently surpassed holistic and histogram features. Despite their extensive use in literature, edge and texture traits did not deliver ideal performance for a holistic approach. As a result, further data sets must be examined to corroborate these findings and to see if other feature sets or regression models might boost performance.

Later, Saleh et al. [31] considered crowd density and visual surveillance to be the most significant aspects in the computer vision research context. The survey focused on two approaches: direct (i.e., object-based target detection) and indirect (e.g., pixel-based, texture-based, and corner points based analysis). As long as people were adequately segregated, direct approaches tracked and counted people simultaneously. The indirect technique, on the other hand, used a collection of measuring features and crowd learning algorithms to conduct the counting and estimating processes. While concluding the direct crowd estimating approach, the survey highlights that in lower-density groups, recognising individuals is easier. When detecting people in large groups or in the presence of occlusions, however, this process became more challenging and complex. That's why despite recent breakthroughs in computer vision and pattern recognition algorithms, many recent studies have avoided the task of detecting individuals to save processing time. Instead, majority of the research has focused on indirect crowd estimation approaches based on a learning mapping between a set of observable variables and the number of people.

Zitouni et al. [32] attempted to provide an explanation of such challenges by extrapolating relevant statistical evidence from the literature and making recommendations for focusing on the general elements of approaches rather than any specific algorithm. The study focused on existing crowd modeling approaches from the literature, concluding that the methods are still density dependent. In addition, real-world applications in surveillance, behavioral understanding, and other areas necessitate that crowd analysis that begins at the macro-level and branches into the micro-level. Let us consider the case of a crowd splitting due to an individual target crossing. Although macro-analysis (in this case, splitting) could detect changes in crowd behavior, micro-analysis (individual target crossing) is required to understand the cause of the behavior. To meet such realistic expectations, crowd representation and inference must concentrate on development at both macro and micro levels as well as in the middle. Most techniques, according to the study, operate under strong and restrictive assumptions such as camera perspective, environmental conditions, density, background and occlusion which must be addressed in the future.

Grant et al. [33] investigated crowd analysis in relation to two main research areas: crowd statistics and behavior analysis. To address the challenge of measuring large crowds with high densities, the survey determined that good data including photographs collected at a variety of standoffs, angles, and resolutions as well as ground-truth labels for comparisons, is essential. It also shed light on the intriguing topic of detecting crowd demographics, where knowing demographics like gender, ethnicity, and age could be beneficial for event planning and marketing. The study also indicated that combining behavior recognition could help determine factors like the quantity of persons walking versus sprinting in a scene. It was strongly stated that synthetic crowd videos filled many gaps, and that these videos were useful in generating important ground-truth information for evaluating and comparing algorithms as well as providing scenes of situations that are too dangerous to re-enact. Plus, it justified the need to generate synthetic crowd data set in the future to avoid such scenarios.

Sindagi et al. [34] compared and contrasted several single-image crowd counting pioneering methodologies and density estimation methods that used hand-crafted representations, with a strong emphasis on newer CNN-based approaches. Across all the data sets, the most recent CNN-based algorithms outperformed the traditional approaches, according to the study. While CNN-based methods performed well in high-density crowds with a variety of scene conditions, traditional approaches had substantial error rates in these situations. Additionally, the multicolumn CNN architecture [10] was tested on three diverse data sets such as UCSD, WorldExpo '10, and ShanghaiTech and the method attained state-of-the-art results on all the three data sets. The CNN-boosting approach by Walach and Wolf [35] achieved the best results on the Mall data set. Optimum results on the UCF_CC_50 data set were achieved by joint local and global count approach [36] and Hydra-CNN [37].

Kang et al. [38] examined crowd density maps created using various methodologies on a variety of crowd analysis tasks, such as counting, detection, and tracking. While fully-convolutional neural networks (e.g., MCNN) produce reduced-resolution density maps performed well at counting, their accuracy decreased at localisation tasks due to the loss of spatial resolution, which cannot be entirely recovered using upsampling and skip connections. It was also recommended that dense pixel-prediction of a full resolution density map using CNN-pixel generated the best density map for localisation tasks, with a minor decrease in counting tasks. Dense prediction, on the other hand, had a larger computational complexity than fully-convolutional networks.

Tripathi et al. [39] offered a thorough overview of contemporary convolution neural network (CNN)-based crowd behavior analysis approaches. The goal of the approaches that were examined was to give law enforcement agencies a real-time and accurate visual monitoring of a busy area. The study identified a shortage of training data sets as a major difficulty when utilising CNN to analyze distinct population types. A list of numerous data sets was offered in this survey. These data sets, however, only comprised a few hundred training examples that were insufficient to train a CNN. CNN-based methods require a large pool of labeled training data sets and major manual interventions that were both complex and time-consuming. Another study found that CNN-based approaches require specialized hardware for training, such as GPUs, because training a CNN is a computationally expensive proposition. To overcome this issue, it would be interesting to look into transfer learning approaches that used previously taught models rather than having to train the model from scratch. Because a shortage of training examples for various types of crowd can impair the system's performance, online CNN training could become an exciting research domain.

Most recently, Gao et al. [40] presented a review of over 220 methods that looked at crowd counting models, primarily CNN-based density map estimates from a variety of angles, including network design and learning paradigms. It tested various state-of-the-art approaches and benchmarked crowd counting algorithms against several crowd data sets such as the National Forum for Urban Research, UCSD, Mall, WorldExpo'10, SHA and UCF-QNRF. The study suggested that PGCNet [41], S-DCNet [42] and PaDnet [43] methods outperformed on Shanghai Tech data set with a MAE of 57.0%, 58.3% and 59.2%, respectively. The study demonstrated, however, that mainstream models were intended for domain-specific applications. Furthermore, supervised learning necessitates precise annotations could be time-consuming to manually label data, especially in highly congested scenarios. Given the unanticipated domain gap, generalising the training model to unseen scenarios might provide sub-optimal outcomes. The study also found that MCNN's [10] head size is proportional to the distance between two people. This notion prompted the creation of a geometry-adaptive kernel-based density map creation method, which has inspired many crowd estimation works to use this tool to prepare training data.

The studies [34,38–40] found that CNN methods are successful and outperform traditional approaches in high-density crowds with a variety of scene variables, whereas traditional approaches suffer from high-error margins in such settings. Sindagi et al. [34] compared different methods for single-image crowd counts and density estimation, the multicolumn CNN architecture [10] performed best on the data sets from UCSD, WorldExpo 10, and ShanghaiTech. Another study from Kang et al. [38] found that MCNN-generated reduced-resolution density maps performed well in crowd counting. Tripathi et al. [39] highlighted a shortage of training data sets as a major issue in utilising CNN to analyze diverse crowd types. It indicated that the existing data sets only had a limited amount of training examples, which were insufficient to train a CNN. Hence, it validated the need to create more training data. After examining more than 220 works, which primarily included CNN-based density map estimation methods, the most recent study from Gao et al. [40] highlighted that MCNN [10] performed well in dense crowd scenes. Considering the suggestions from the most recent studies, we have applied the MCNN and used ShanghaiTech data set to train, test, and validate the simulation system against real-crowd data and have discussed it further in Section 3.4.

2.4. Related Open Source Crowd Simulator

Considering the recent growth in crowd estimation, testbed for generating crowd training and testing data is a major issue. Capturing the crowd has never been easy and ethical issues don't allow to capture the crowd in most countries. Furthermore, the process of capturing the crowd is expensive and can backfire. Simulators are the best solution to overcome the testbed issue because they are cost-effective and can easily produce data for training and testing. To resolve the issue, we have reviewed the existing crowd simulators, determined its limitations, and demonstrated why a new 3D crowd simulation system with an integrated UAV simulator is required.

Kleinmeier et al. [44] introduced Vadere, a framework for simulation of crowd dynamics. It consists of features that allowed interaction with the microscopic pedestrian. As a result, it has contributed to many simulation models and comprises of models like the gradient navigation model and social force model for further research purposes, which are restricted to 2D simulations. Maury et al. [45] introduced Cromosim, a library specifically designed for Python that was mainly used to model crowd movements. It is simple to set up, and there are some examples models available to monitor the trend, such as follow-the-leader and social powers. However, its use is limited in other respects, such as crowd motion tracking. Curtis et al. [46] developed Menge, a full-featured 3D crowd simulator designed for crowd simulation and dynamics that compared two different models. Since the crowd and its aspects do not appear to be real, it could only be used for tracking purposes within the developed environment.

Crowd Dynamics is another 2D simulation system intended to develop for crowd movement. However, the system is still in the early stages of development. PEDSIM is a microscopic pedestrian crowd library with limited application. Consequently, the documents simply mention the use of PEDSIM to implement several models such as cellular automata and social force, but nothing else is specified. Wagoum et al. [47] presented JUPEDSIM framework to map crowd dynamics. The framework is an open source one and can be used for research purposes such as mapping and measuring crowd dynamics, data visualization etc. Mesa [48] is a python library limited to modeling functions and can't be used for simulation. There isn't enough data available for RVO2 [49], Fire Dynamics [50] and AgentJL. These frameworks have been developed specifically for crowd dynamics navigation and haven't been updated in a long while. Other licensed and paid crowd simulators such as CrowdSim3D and Urho3D are also available with built-in tracking and mapping features, but their use is limited and they are expensive.

To summarise, except for Menge, most open-source simulators are limited to 2D and are specifically designed to track the crowd dynamics and motion. A detailed review of the available open-source simulators summarised in Table 1 has revealed that most simulators are designed for specific tasks such as crowd dynamics or fire dynamics study in 2D and are not efficient enough to generate 3D synthetic data and avatars to mimic real-world conditions. Considering the available 3D simulator 'Menge', it consists of repeated characters and encounters problems in distinguishing various features of individuals such as gender, age, weight, height, ethnicity, proportion, outfit, pose, color, and geometry. As a result, this 3D simulator cannot be used to envision any scenario that mimics real-world settings. Furthermore, various geometric shapes and topologies for each individual's eyes, hair, teeth, eyebrows, eyelashes, and other features are necessary to produce a realistic prototype. Menge does not provide this functionality. All these factors validate the need for developing a new 3D crowd simulator that can generate reams of data in any scenario, visualise a realistic 3D world, as well as relative locations for crowd estimation.

Table 1. Summary of Open-Source Simulators and Supporting Libraries.

Simulator	Language	OS	2D/3D	Intended Use
Vadere [44]	Java	Windows, Linux	2D	Crowd Dynamics
Cromosim [45]	Python Library	Windows	2D	Crowd Motion
Menge [46]	C++	Windows, Linux	Both	Cross-Platform
Crowd Dynamics	Python	Ubuntu 16.04	2D	-
PEDSIM	C++	Win, Linux	2D	Pedestrian Library
JuPedSim [47]	Python, C++	Windows, Linux	2D	Pedestrian Dynamics
Mesa [48]	Java	Windows, JVM	2D	Crowd Simulation
RVO2 [49]	C++	-	2D	Mobile Robots
Fire Dynamics [50]	-	Windows, MacOS	2D	Fire Dynamics
Agent.JL	Julia	-	2D	Agent-Based Model

2.5. Crowd Data Sets

Traditional surveillance systems for crowd estimation are effective when dealing with small crowd sizes. Nevertheless, the traditional approach has some design issues including slow frame processing speeds, resulting in a major breakdown in the process because it cannot handle high-density crowds. Most of the methods have been developed and tested for single images or videos, with majority of the approaches perform crowd testing with low-density crowds [2]. This study analyzes crowd data sets and subclassifies them into free, surveillance and drone-view crowds. The data sets have been categorized based on release year, attribute, number of samples, and average count per image. The primary objective of this study is to identify why existing drone view data sets cannot be used for the estimation of crowds using UAV, and why synthetic data is required.

The first free-view data sets UCF_CC_50 [51] were released in 2013 with a sample size of 50 and 63,974 instances. UCF_CC_50 is the only available large density crowd data set as shown in Table 2. ShanghaiTech Part A [10] is another congested attributed data set containing 241,677 instances with an average count of 501 people. Sindagi et al. [52] discussed available data sets for crowd surveillance and estimations. Some of the popular and easily accessible data sets include UCSD [26] which consists of 2000 frames of size 238×158, and the Mall data set [29] containing 2000 frames of size 320×240 with 6000 instances and large number of labeled pedestrians. The ShanghaiTech crowd data set [10] discussed in Table 3 includes both part A and part B of the dataset. It consists of 1198 images with a large number of 330,000 annotated heads.

Table 2. Summary of Different Free-View Crowd Data Sets.

Data Set	Year	Attribute	No. Samples	No. Instances	Avg. Count
NWPU-Crowd [53]	2020	Localization	5109	2,133,375	418
JHU-CROWD++ [54]	2020	Congested	4372	1,515,005	346
UCF-QNRF [55]	2018	Congested	1535	1,251,642	815
SanghaiTech Part A [10]	2016	Congested	482	241,677	501
UCF_CC_50 [51]	2013	Congested	50	63,974	1279

Bahmanyar et al. [66] presented the first drone-view crowd data set in 2019 for crowd estimation known as DLR's Aerial Crowd Data Set. The images were captured through a helicopter providing 33 aerial images from 16 different fights of a slowly moving crowd. Zhu et al. [67] presented the second aerial data set of the crowd. As shown in Table 4, this data set comprised of 112 video clips collected from 70 different scenarios.

Table 3. Summary of Different Surveillance-View Crowd Data Sets.

Data Set	Year	Attribute	No. Samples	No. Instances	Avg. Count
DISCO [56]	2020	Audiovisual	1935	170,270	88
Crowd Surveillance [41]	2019	Free scenes	13,945	386,513	28
ShanghaiTechRGBD [57]	2019	Depth	-	-	-
Fudan-ShanghaiTech [58]	2019	Video	15,000	394,081	27
GCC [59]	2019	400 Fixed Scenes	15,211	7,625,843	501
Venice [60]	2019	4 Fixed Scenes	167	-	-
CityStreet [61]	2019	Multiview	500	-	-
Beijing-BRT [62]	2019	1 Fixed Scene	1280	16,795	13
SmartCity [63]	2018	-	50	369	7
CityUHK-X [61]	2017	55 Fixed Scenes	3191	106,783	33
ShanghaiTech Part B [10]	2016	Free Scenes	716	88,488	123
AHU-Crowd [64]	2016	-	107	45,000	421
WorldExpo'10 [65]	2015	108 Fixed Scenes	3980	199,923	50
Mall [29]	2012	1 Fixed Scene	2000	62,325	31
UCSD [26]	2008	1 Fixed Scene	2000	49,885	25

Table 4. Summary of Different Drone-View Crowd Data Sets.

Data Set	Year	Attribute	No. Samples	No. Instances	Avg. Count
DroneVehicle [68]	2020	Vehicle	31,064	441,642	14.2
DroneCrowd [67]	2019	Video	33,600	4,864,280	145
DLR-ACD [66]	2019	1 Fixed Scene	33	226,291	6857

According to the pattern since 2008, when UCSD's first crowd data set was released, the majority of publicly available crowd data sets have been captured with static cameras [34] and have been limited to 2D. The first aerial crowd data set was released in 2019 with a sample size of 33. That said, the data set is inaccessible and has no annotations. In fact, most aerial crowd data sets are not widely available for study. Previous studies [39] evidenced and highlighted the shortage of training examples for various crowd types. Ref. [20] focused primarily on simulation realism and highlighted the importance of virtual data sets that will address the issue in the near future. Our study also concluded that gathering and manually annotating crowd data sets are both expensive and time-consuming. Considering the current laws and ethical issues, there is a justified need of a testbed that can generate virtual crowd data set and contain in-depth information of both 2D and 3D images.

The study of various traditional and most recent 2D crowd estimation approaches discussed in Sections 2.2 and 2.3 highlighted the inherent limitations of 2D approaches which include static camera monitoring that can monitor a specific area with a high possibility of counting the same individual multiple times, nonhuman or false detections, and lack of information and inconsistency in performance, which leads to a high percentage of errors, among others. Extensive work has been done for different segments of 2D crowd estimation and received a lot of attention, but work related to 3D crowd estimation is limited. Recently, promising research has been conducted on density estimation [69–72] and advances have been made in 3D pose estimation from 2D [11,73–80] but the work related to 3D crowd estimation using UAV is not prevalent. Interestingly, UAVs have immense potential to estimate the crowd spread over a huge geographical area in a shorter duration. Recent advances in optical instrumentation and computer vision techniques have facilitated the use of 3D models to explore in-depth information. In contrast, very little research has been done regarding 3D crowd estimations with UAVs. This fact alone underscores the necessity to devise a new way to overcome the traditional and inherent limitations. It also demonstrates how our 3D work varies from existing and conventional 2D crowd estimation methodologies. To summarise, taking into account the existing shortcomings, we have presented a 3D crowd simulator in Section 3.3, trained, tested and validated the simulation system against real-crowd data in Section 3.4. In addition, we have introduced a novel 3D crowd estimation technique using UAV for a robust and accurate estimation of a crowd spread over a large geographical area in the subsequent sections.

3. 3D Crowd Estimation Using UAV

In this section, various techniques and tools used to develop 3D crowd estimation technique with UAVs have been covered in detail. We have also highlighted the way these tools can be used in conjunction with one another. An overview of the development of a crowd simulation for training and testing data has also been discussed. Unreal Engine has been used as the main tool for simulation and Make Human and Anima have been employed to design and import random crowd that mimic real-life settings. Furthermore, we have discussed the process used for training, testing and validation of synthetic data against real crowd data and vice-versa. Finally, we have introduced our novel method of 3D crowd estimation using UAVs in real time.

3.1. 3D Simulation and Modeling

Unreal Engine [81] is a game engine developed by Epic Games that focuses on first person shooter games. It was created using Blueprint and C++ as the main languages in version 4 (v4). With features such as blueprint interface, game mode, simulation, real-time output and automatic annotations, it is the perfect fit for reproducing 3D framework, especially for simulating real-life scenarios that rarely occur.

Make Human [82] is an open-source 3D computer graphics software used to create realistic humanoids. Make Human is used to design and create crowds size considering different genders, age and features. Given a larger community comprising of programmers, artists and people with academic interests in 3D modeling of characters, this tool is written in Python and is compatible with almost all the available operating systems. Make human is easy to use and extracts the skeleton or a static mesh as per the requirements of any other simulation tool such as unreal engine.

Anima [83] is a 3D people animation application developed specifically for architects and designers and is ideal for creating amazing 3D animated people quickly and easily. The tool has been used to create many 3D animated people and realistic scenarios. The crowd flow and movement direction are plotted in such a way to avoid collision and maintain a realistic flow. Many realistic 3D models such as stairs, escalators, tracks, and moving sidewalks are pre-designed and easy to access for UE4 which not only helps to design and simulate any complex scenarios quickly but also saves time while creating any new realistic setting.

Colmap [84] is a 3D reconstruction tool and uses the patch-based stereo to reconstruct 3D dense point clouds. In our proposed method, it has been used to generate 3D models using images extracted from Unreal Engine. The Colmap provides intrinsic parameters such as camera model and extrinsic parameters such as camera location, rotation, etc. Several studies [85,86] critically compared the results of popular multiview stereo (MVS) techniques and concluded that COLMAP achieves the best completeness and on average, it produced promising results for most individual categories.

3.2. Why 3D Simulation?

3D simulation is less time-consuming and cost-effective to build a 3D simulator of a crowd to train and test the system as well as provide accurate ground truth information about people and their locations. In addition, a 3D simulator is useful to create and design simulations of seldom occurring events and understanding the real-world outcomes. Additionally, it can be useful to self-train the system by finding out about those uneven possibilities, such as stampedes, public gatherings, etc. within the crowd data sets. Another factor that influences and attracts computer vision researchers toward 3D simulation is the virtual data set. It makes it possible to consider and construct a virtual data set by creating various scenarios, events, and their outcomes in real time, which can help to train and test the system [87].

3.3. Overview of the Proposed Testbed

In a limited time, UAVs have gained enormous prominence due to their ability to resolve major issues. Obtaining a licence and permission to fly a UAV near a crowd in most countries is hard, expensive, and time-consuming due to rigorous restrictions and regulatory limits. Navigating and coping with a variety of precise settings and unforeseen situations can also be difficult. Handling a UAV in a gusty environment with a shorter flying time and distance, for example, highlights the inefficiency of mapping a large area, which could be dangerous in real life. All of these variables make 3D simulation the ideal solution because it has no negative implications or ethical issues.

Considering the challenges of gathering and annotating real data, we built the crowd simulation system using the Unreal Engine version 4 (UEv4). The design of the basic prototypes and reusable meshes such as houses and trees was the first step involved in creating a virtual environment and shown in Figure 1. Furthermore, we placed all those meshes within the environment to give a real-life look. We have used smooth, linear, and spherical features to flatten and reduce surface noise. Animation and wind effects were incorporated to make the virtual environment more realistic, but only in 3D. These models can be imported and utilized in a variety of settings, making the process of building and generating scenarios quick, easy, and adaptive to the requirement.

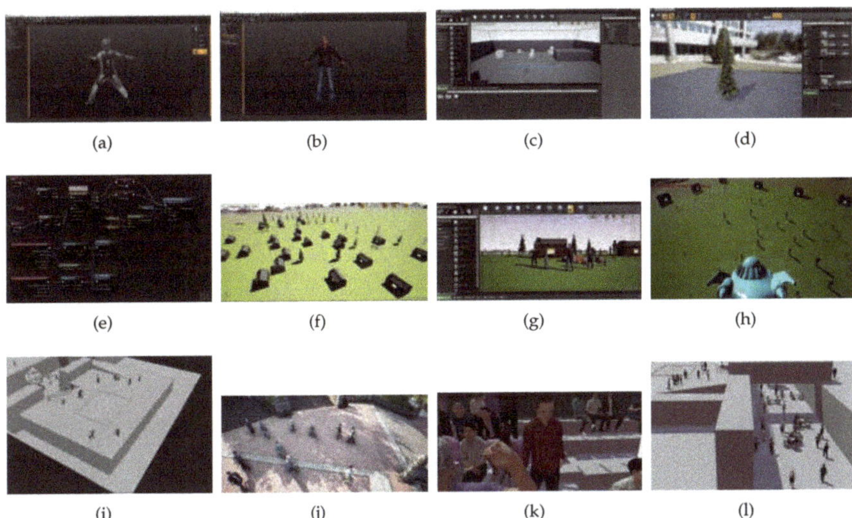

Figure 1. The figure demonstrates various steps involved before simulating the crowd, whereas (**a**) shows the skeleton design sample of a person containing in-depth details which can be exported and further used in UE4, (**b**) shows one of the samples designed to be a part of the crowd, (**c**) contains the basic template for the first person used as a map to place various objects, (**d**) demonstrates the designing and animating of the mesh. A tree sample has been presented in this image. (**e**) shows a sample blueprint command line to calibrate and establish the working between different objects, (**f**) shows an initial output map designed before placing the crowd in the environment, (**g**) demonstrates how different crowd samples look like when they are ready for simulation, (**h**) depicts the final image after starting the simulation where the image was captured from the top view and showed our UAV prototype used in UE4, (**i**–**l**) demonstrates various scenarios where the crowd was randomly distributed in diverse settings.

Having said that, it is necessary to create a synthetic crowd prototype comprised of different genders before simulating the environment. Hence, we have used Make Human and Anima to design and generate random crowds using random sets of features for different random variables to mimic the real world. The random crowd consists of individuals of different genders, ages, weights, heights, ethnicities, proportions, outfits, poses, colors, and geometries. For proper representation, we have used different geometric shapes and topologies for the eyes, hair, teeth, eyebrows, eyelashes, etc. of each individual.

Manually annotating the crowd in any dense crowded image is an extremely laborious and time-consuming task with a higher possibility of getting false annotations or multiple count of the same individual. While the captured 2D data holds good image resolution, the inherent limitation of 2D does not make it efficient to provide every single detail required for estimating the crowd in 3D. That said, data collection within the 3D simulation system is relatively easy and accurate to generate reams of data, especially when using a moving camera over a large crowd. Our proposed 3D simulation system is efficient enough to generate automatic annotations and can provide 3D world and relative locations to estimate the crowd in any static or dynamic event. The simulation system is also able to generate virtual data sets that could be beneficial in future research within the domain. Furthermore, it resolves existing issues such as the availability of massive crowd data sets, among others. The flowchart in Figure 2 depicts the steps taken to capture the frame while storing ground truth (GT) positions at the same time. The collected data was used in the subsequent Section 3.4 for the training, testing and validation of the simulation system and generation of synthetic data. Furthermore, the 3D annotations collected by flying the UAV were extracted from the simulator and further used in the final 3D method introduced in Section 3.5.

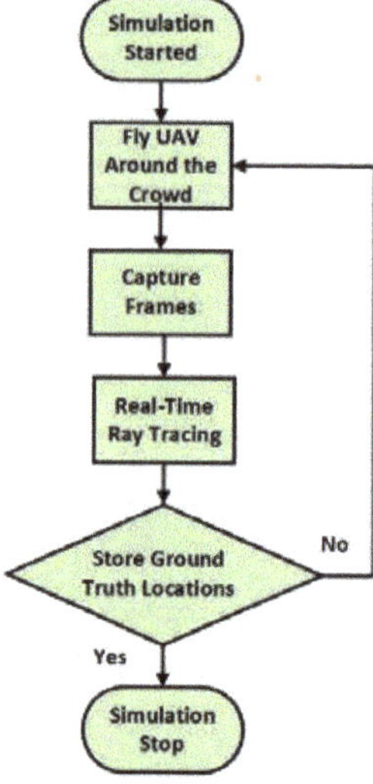

Figure 2. The flowchart presents the whole pipeline for capturing synthetic data with a 3D simulation system.

3.4. Training, Testing and Validation the Simulation System against Real-Crowd Data

Given the requirement for a UAV crowd estimation and limitations of a flying UAV in the real world, we have introduced a novel way to estimate the crowd using synthetic images extracted from our simulation system. We started by implementing our initial idea of building and testing the simulation system. Using aerial photos gathered from the drone as a foundation for assessment was a very challenging task. So we prototyped photo-realistic humanoids of various sizes and integrated all of their meshes and skeletons into the simulator to make it as realistic as possible.

With a variety of methods discussed in Section 2.3, we have evaluated the advantages and disadvantages of the broad approaches. Most recent studies [34,38–40] suggest that multicolumn CNN [10] method achieves the best results on ShanghaiTech data set and is efficient enough to train, test and validate the simulation system against real-crowd data. ShanghaiTech is the best fit as it is one of the largest large-scale crowd counting data sets in previous few years. It consists of 1198 images with 330,165 annotations. According to different density distributions, the data set has been divided into two parts: Part A (SHA) and Part B (SHB). SHA contains images randomly selected from the internet, whereas Part B includes images taken from a busy street of a metropolitan area in Shanghai. The density in Part A is much larger than that in Part B which make SHA a more challenging data set and an ideal fit for large crowd testing.

To test and validate the simulation system, we extracted the aerial video captured through UAV within the simulator and split it into different frames. Initially, we set up the ShanghaiTech data set for testing and validation against the synthetic images (Figure 3). For testing the system, we set up data and created the training and validation set along with ground truth files. We calculated the errors using mean absolute error (MAE) and root mean square error (RMSE), and the output in the form of density maps.

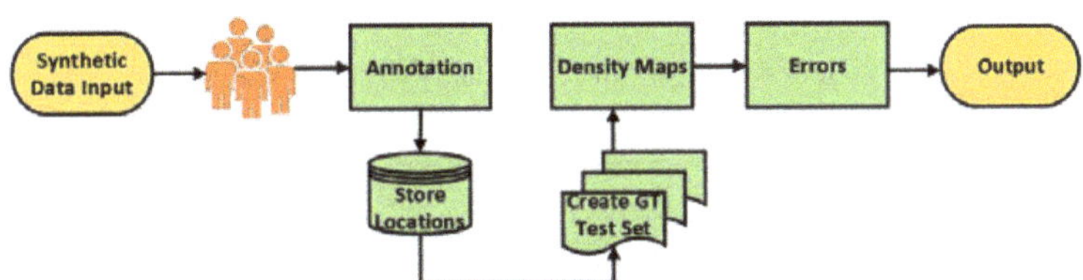

Figure 3. The pipeline shows the steps involved in testing of synthetic data against publically available crowd data set.

We trained the model on synthetic data using multicolumn convolutional neural network (MCNN) after obtaining a high throughput and validating the simulated data. Three parallel CNNs, whose filters were attached with local receptive fields of different sizes, were used as shown in Figure 4. We utilized the same network structures for all the columns (i.e., conv–pooling–conv–pooling) except for the sizes and numbers of filters. Max pooling was applied to each 2×2 region, and Rectified linear unit (ReLU) was adopted as the activation function. We used fewer filters to minimise computation time.

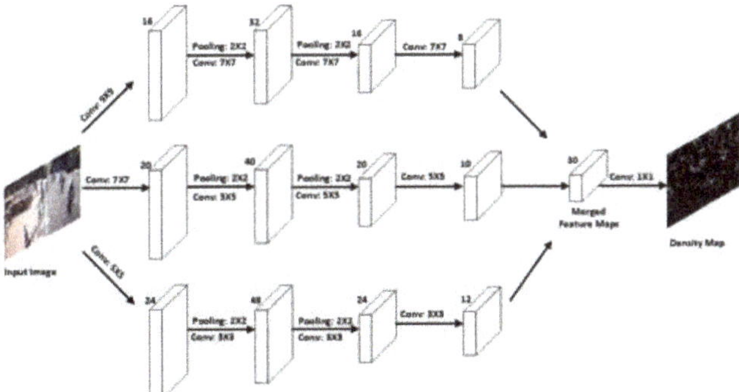

Figure 4. The figure depicts the network architecture design and overview of single image crowd counting via multi-column network.

3.5. Our Approach to Crowd Estimation Using UAV

In this section, various tools and techniques used to develop the 3D crowd estimation technique using UAV have been discussed. We have highlighted step-by-step how these tools are interlinked with each other. We have briefly discussed the Make Human and Anima for designing and importing random crowd that mimic real-life settings.

In the most recent studies, counting the same individual from a moving camera has been a major issue. We have attempted to overcome the issue by introducing a novel 3D crowd estimation technique using UAV for a robust and accurate estimation of a crowd spread over a large geographical area. Figure 5 shows the step-by-step process of our presented method where the basic prototypes and meshes were designed to setup a simulation environment in Unreal Engine. Anima and Make Human were used to generate random crowds size using random sets of features for different random variables to mimic real-life settings. After preparing the simulation environment, we flew a virtual UAV around the crowd and captured the ground truth 3D locations which we will use at the end to map the estimated 3D crowd locations. Various frames were also captured associated with the crowd to train, test and validate the system. After extracting the captured data from Unreal Engine, we tested the captured virtual data using state-of-the-art method MCNN. Later, Laplacian of Gaussian (LOG) was applied in the extraction of the density map provided by the MCNN to identify the possible 2D crowd location. It was later used to ray trace the possible crowd locations in 3D. In the third step, we reconstructed the 3D model from the frames captured using UE4 and collected in-depth details of the model such as camera location, quaternion matrix, camera translation and points such as screen points and 3D points for every 2D image provided as input. Finally, we initiated a ray hit testing and traced the possible 2D crowd location extracted from the blob-detector and stored the intersection points between ray and plane, considering them as the possible crowd locations in 3D model. Although the traced 3D locations overlapped in the initial frame capturing, we set up an averaging method and discarded most of the overlapping points from each frame. To map the output estimated point with the ground truth point captured from UE4, we used the ICP algorithm for registering both point sets. Once it converged, we mapped the ground truth points with the estimated points using the nearest neighbour search algorithm and extracted the matched pair between the two sets, where p_i is considered a match to q_j if the closest point in Q to p_i is q_j and the closest point in P to q_j is p_i and tested it against various universally-agreed and popularly adopted measures for crowd counting model evaluation which have been discussed further in Section 3.6.

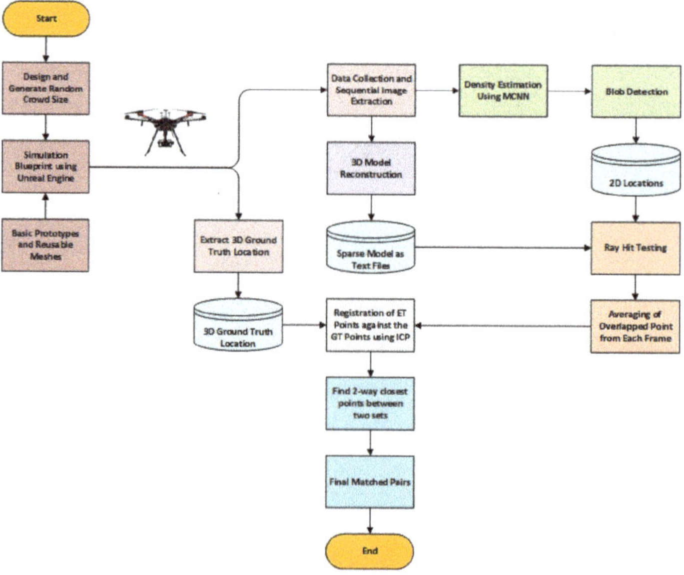

Figure 5. The system architecture diagram provides a detail representation and steps involved in our approach to crowd estimation using UAV.

To make a clearer representation of the method, we divided the process into several steps and tried to present the working of every step. To give a realistic vision, we also attempted to visualise how the output would look like. The steps involved in the presented method are as follows:

Step 1: Make Human and Anima have been used to design and generate random crowds size using random sets of features for different random variables. They give a random crowd that mimics real-life settings. Furthermore, this image covers people of different genders, age, weight, height, ethnicity, proportion, clothes, pose, colour, geometries etc. Different geometries for each person have been used to make the synthetic crowd more appropriate for the real crowd, including eyes, hair, teeth, topologies, eyebrows, eyelashes, and so on. Furthermore, these individuals were involved in the simulation system's estimation process.

Step 2: Unreal Engine (UE4) is primarily used as a platform to simulate various real-life scenarios that rarely occur. To make it more practical and closer to real-life situations, we have used random crowd distribution (Figure 6). Because of the random distribution, the crowd size for each simulated scenario is unknown before testing. Algorithm 1 demonstrates steps 1 and 2 with a detailed overview of how the simulation scenario was created and 3D locations were extracted for the simulated crowd within the system.

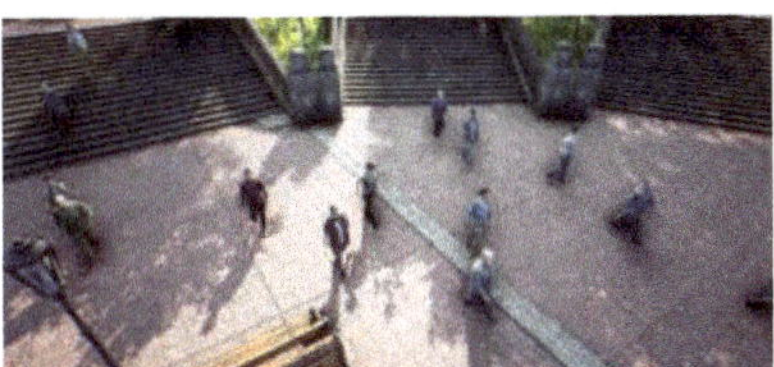

Figure 6. The figure shows the demonstrations from Steps (1–2), where the synthetic image has been captured from the UAV.

Algorithm 1 Algorithm for 3D simulation and data collection.

Input: $H=\{h_1, h_2, ..h_{126}\}$, Where H is a set of humanoid;
$G=\{g_1, ..g_{1000}\}$, where G is a set of environmental geometries
Output: Simulation Scenario S;
Log file LF consisting of 3D locations P_i;
Frames captured F_i
while *Not enough sample picked* **do**
    /*enough sampled here means the no. of humanoid objects required for
    simulation*/;
    Import a random h_i
end
while *Not enough sample picked* **do**
    /*enough sampled here means the no. of geometric objects required for
    simulation*/;
    Import a random g_i;
end
Combine the imported subsets C, which represents the crowd and M the mesh
  respectively to get the simulation scenario S;
Start simulation;
while *Not all crowd captured* **do**
    F_i=fly UAV around the crowd to capture frames;
    Ray tracing ;
    L_i=extract crowd 3D locations P_i to log file LF;
    Store locations in log file LF ;
end
return LF containing GT location P_i;

Step 3: Density estimation and blob detection were used for projection and verification. To gather all the information and evaluate the output images, the system was trained using real images and tested on synthetic images provided by the simulator. Few state-of-the-art pre-trained models were considered for checking against both the synthetic and real data to train and test the system. Moreover, we incorporated a multicolumn convolutional neural network for single image crowd counting. We repeated this process for all the data. The density heat maps generated using the person detector (Figure 7) for all the 2D images were used for mapping the 3D data. Later, we used a Gaussian blob detector to extract the individual's 2D locations from the density maps. The coordinates were later used to ray trace these 2D locations to obtain the 3D locations. These points were crucial for filtration and determining whether or not the estimate point in the 3D model belonged to a person. Algorithm 2 demonstrates step 3 and highlights the procedure followed to extract the 2D coordinates for each person from each image that has been extracted.

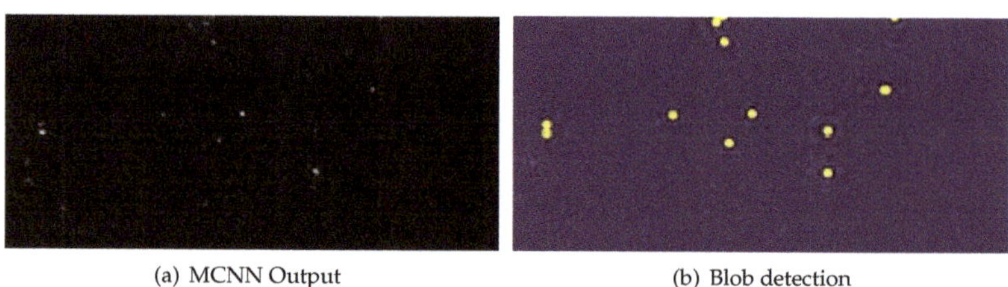

(a) MCNN Output   (b) Blob detection

Figure 7. The figure shows the demonstrations of step 3, whereas (**a**) shows the network output from MCNN in the form of density map and (**b**) represents the Step 3, where the blob detected from the density map are shown and further used for mapping and tracing the crowd.

Algorithm 2 Algorithm for density estimation and Blob detection.

Input: Frames captured F_i
Output: .JSON File containing 2D coordinates (x_i, y_i) for each person EP_i in
      frame F_i;
Initialization;
Download data set;
create directory;
Density_map = MCNN(F_i);//use MCNN Algorithm here to get heat map
for $EP_i \leftarrow 0$ **to** F_i **do**
    Read density_map to array;
    Convert to gray scale;
    Apply Laplacian of Gaussian in image;
    Detect blobs;
    Extract (x_i, y_i);
    Save extracted (x_i, y_i) to .JSON File
end
return .JSON File;

Step 4: Colmap is used to generate 3D models using the synthetic data (Figure 8) gathered from the simulator. Various simulated images were captured by flying the UAV over the randomly distributed crowd. The gathered data was merged into a realistic model using structure-from-motion (SfM) and multiview stereo (MVS). The whole pipeline returned the 3D parameters such as camera location, quaternion matrix, camera translation and points such as screen points and 3D points for every 2D image provided as input.

Figure 8. The image presents the first step of COLMAP 3D reconstruction where a set of simulated overlapped images have been provided as an input.

This approach uses a set of multiview images captured by RGB cameras to reconstruct a 3D model from the object of interest. 3D reconstruction is often identified as SfM-MVS. SfM is an acronym for structure-from-motion. It creates a sparse point cloud model from the input images.

First, the SfM technique determines intrinsic (distortion, focal length, etc) and extrinsic (position and orientation) camera parameters (Figure 9) for putting the multiview images into context by identifying the local features/keypoints of the images. The corresponding points were then used to measure the 3D model and find the relationship between images. Algorithm 3 represents how the 3D model (Figure 10) has been reconstructed as explained in step 4.

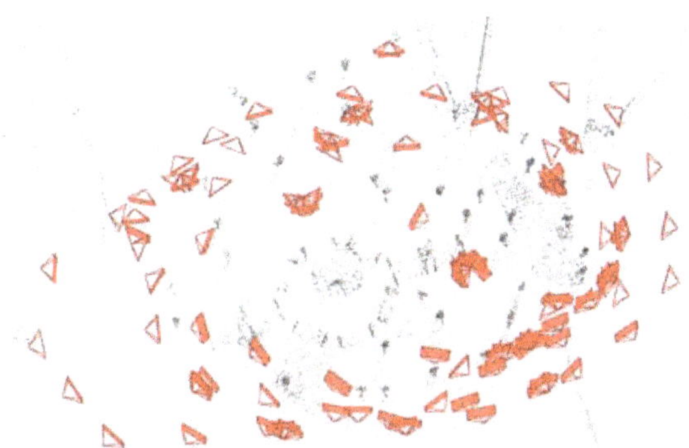

Figure 9. The figure shows the UAV path trajectory. The data was captured by following a circular path to store every crowd detail from the scene.

Figure 10. The figure explains the final step involved in the COLMAP reconstruction. A 3D model has been provided as an output.

Algorithm 3 Algorithm for 3D model reconstruction using COLMAP.

Input: Frames captured F_i
Output: Reconstructed 3D Model as STL file SF;
Cameras File CF;
Images File IF;
Point 3D File PF;
Initialization;
while *3D model not reconstructed* **do**
    | Feature detection and extraction;
    | Feature matching and geometric verification;
    | Structure and motion reconstruction;
end
return model SF as text to store the values of CF, IF,PF

Step 5: A ray hit test was set up using the starting point and direction to find the intersection point between the ray and 3D model plane. It was used to track down and estimate the crowd size in 3D, while considering the challenge of a moving camera and crowd. It is possible to ray trace every point in each 2D image, but it would be a very expensive and time-consuming process. To overcome this problem, unrelated points were filtered and discarded and ray trace was set up only for the points extracted after the blob detection obtained the exact 3D location points. The returned ray intersection points with the relevant frame numbers were stored and used in the next step to overcome the issue of counting the same individual multiple times.

Structure-from-Motion (SfM) is the process of reconstructing 3D structure from its projections into a series of images. The input is a set of overlapping images of the same object taken from different viewpoints. The output is a 3D reconstruction of the object as well as the reconstructed intrinsic and extrinsic camera parameters of all images. Typically, Structure-from-Motion systems divides this process into three stages: feature detection and extraction, feature matching and geometric verification and structure and motion reconstruction. Furthermore, multiview stereo (MVS) takes the SfM output to compute depth and normal information for every pixel in an image. Fusion of the depth and normal maps of multiple images in 3D then produces a dense point cloud of the scene. Using the depth and normal information of the fused point cloud, algorithms such as the Poisson surface reconstruction [88] can then recover the 3D surface geometry of the scene.

Figure 11 depicts the original model reconstructed using an overlapped image provided as input. Before moving forward, plotting the traced point back is an efficient way of checking the accuracy. For this, we used the reconstructed intrinsic and extrinsic camera parameters of all images stored in a database. Later, we plotted the same traced points back to create the same model to double-check the data accuracy. Figure 12 refers to the back projected traced points to the point cloud which creates an accurate model and proves the reconstructed model's accuracy. Various steps followed in step 5 have been presented in Algorithm 4 that demonstrate how the intersection points were extracted using a ray hit test.

Step 6: A merging algorithm was developed to find the average of the total number of points hit by the ray tracer. Then, a list of intersection points for each ID and the threshold was set up as the input. The closest point to the threshold was selected. Each point of the frame number (from ID 1 to N-1) was checked against all the neighbouring points with the same ID. If the difference between the point P and the intersection point Q was greater than the threshold, the point was appended to a new point set while the rest were discarded. The detailed explanation and the steps involved in the algorithm have been discussed in Algorithm 5.

Algorithm 4 Algorithm for Ray Hit Testing.

Input: Cameras File CF;
Images File IF;
Point 3D File PF;
Blob Points 2D File BF;
STL File SF;
Output: Intersection point set p_{ij} for each frame F_i
Initialization;
Transform camera file into key values where key=C_i, camera id;
value=C_p, camera parameters;
Transform image file into key values where key=$P2D_i$, point 2D id;
value=I_p, image parameters;
Transform Point 3D File into key values where key=$P3D_i$, point 3D id;
value=$3DP_i$, 3D parameters;
Map data;
Map Cameras File, Images File, Points3D File, Blob Points 2D File;
Map data output;
for $F_i \leftarrow 0$ **to** $N-1$ **do**
    Map C_i to IF point 2D;
    Extract image data and point data;
    Map image data with blob data;
    Create kD tree from blob points;
    for $P2D_i \leftarrow 0$ **to** $N-1$ **do**
        if *Closest to blob point in kD tree* **then**
            Map with $P3D_i$;
        else
            Eliminate $P2D_i$;
        end
    end
    Caster = rayCaster.fromSTL(STL File, scale=1);
    Read Data cam id, Data parameters, Caster;
    Set up ray start point and direction;
    Intersection points = caster.castRay(start point, direction);
end
return Intersection points p_{ij};

Figure 11. The figure shows the original 3D Model reconstructed using five humanoid prototypes.

Figure 12. The figure has traced points back projected to the point cloud, while reconstructing the original model.

Algorithm 5 Merging of Ray traced intersection points.

Input: List of intersection points p_{ij} for each frame F_i;
threshold t
Output: Pointset Q
Initialisation;
p_{ij} = points[0];
for $id_i \leftarrow 0$ **to** $N-1$ **do**
    new_point_set = Q_i;
    for $p \leftarrow 0$ **to** F_i **do**
        for $q \leftarrow 0$ **to** p_{ij} **do**
            if $(|p-q| > t)$ **then**
                $Q \leftarrow p \cup Q_i$;
        end
    end
end
return Q

Step 7: Point matching for evaluation was carried out in the final step. To evaluate our detections, we had to match the ground truth 3D locations to the estimated locations from our system. The Iterative Closest Point (ICP) [89,90] algorithm was used to find the best fit transform and to validate the estimated points against the ground truth points. Fast Library for Approximate Nearest Neighbors (FLANN) [91] was used for the nearest neighbour search. A two-way matching of points was carried out and cross-checked between the two 3D point sets pairs. Algorithm 6 demonstrates the procedure followed for the two-way matching from the two different point sets where Q represented the estimated average points and P represented the ground truth points extracted from the simulation system. It has been explained in step 1 of the presented method.

Algorithm 6 Algorithm for 2-way point matching using ICP.

Input: $P = p_0, p_1, ..., p_N$;
$Q = q_0, q_1, ..., q_M$;
Output: Matched pairs
Initialize transform M to be the identity;
until converged;
R = Find 2-way closest points between P and MQ $(MQ = Mq_0, Mq_1, ...Mq_M)$;
update M based on matches in R;
return Pairs $(P \rightarrow MQ)$ and $(MQ \rightarrow P)$;

3.6. Evaluation Metrics

Many evaluation metrics are available to predict the estimation and ground truths. They are universally agreed and popularly adopted measures for crowd counting model evaluation. They are classified as image-level for evaluating the counting performance, pixel-level for measuring the density map quality and point-level for assessing the precision of localisation.

The most commonly used metrics include Mean Absolute Error (MAE) and Mean Squared Error (RMSE), which are defined as follows:

$$MAE = \frac{1}{N} \sum_{i=1}^{N} |C_{Ii}^{pred} - C_{Ii}^{gt}| \tag{1}$$

where N is the number of the test images, C_{Ii}^{pred} and C_{Ii}^{gt} represent the prediction results and ground truth, respectively.

$$RMSE = \sqrt{\frac{1}{N} \sum_{i=1}^{N} |C_{Ii}^{pred} - C_{Ii}^{gt}|^2} \tag{2}$$

Roughly speaking, MAE determines the accuracy of the estimates whereas $RMSE$ indicates the robustness of the estimates.

Precision is a good measure to determine when the costs of False Positive are high. For instance, in the current crowd estimation approach, a false positive means that a point hit by the model is not the right point (actual negative) and has been identified as a person (predicted crowd). The crowd estimation system might lose the actual individual out of the crowd, if the precision is not high for the crowd estimation model.

$$Precision = \frac{TruePositives}{TruePositives + FalsePositives} \tag{3}$$

Recall calculates how many of the Actual Positives our model has captured by labeling it as Positive (True Positive). For instance, in the current system, if an individual (Actual Positive) is not predicted and counted null (Predicted Negative), then the cost associated with False Negative will be extremely high, and it might collapse the whole estimation model.

$$Recall = \frac{TruePositives}{TruePositives + FalseNegative} \tag{4}$$

F1 Score may be a better measure to use, if we need to strike a balance between Precision and Recall and see if there is an uneven class distribution (a large number of Actual Negatives).

$$F1 = 2 \times \frac{Precision * Recall}{Precision + Recall} \tag{5}$$

4. Implementation Details

In our experiments, we used Pytorch for training and testing synthetic data. For the hardware equipment, the training was done on a 64-bit computer with 32 cores Intel(R) Xeon(R) Gold 6130 CPU @ 2.10GHz processors, 48 GB RAM and two Tesla P100-PCIE-16GB GPU devices. To improve the training set for training using MCNN, we cropped 9 patches from each image at different locations; each patch was $\frac{1}{4}$ size of the original image. We trained 133 images that contained 1197 patches using the MCNN model. The 2D detector model was trained on a shared network of 2 Convolutional layers with a Parametric Rectified Linear Unit (PReLU) activation function after every layer to enhance the accuracy of the traced blob points. For the CMTL training, we cropped 16 patches from each image at different locations; each patch was compressed to $\frac{1}{4}$ size of the original image. We trained 133 images containing 2128 patches using the CMTL model.

The implementation of 3D crowd estimation was performed using a ray caster on the reconstructed 3D model. The model was reconstructed using the 127 images that were captured from our 3D simulator. The model was rebuilt using an Intel Core i7-8750H processor with a 6 Cores/12 Threads @ 4.1 GHz CPU, Windows 10 on 16 GB RAM, and an NVIDIA Geforce GTX 1060 Max-Q graphics card (6GB of dedicated memory). We used a simple radial camera to capture the data while flying the UAV above the crowd. Initially, the data was captured by following a circular path. The capturing angel varied from $45°$ to $90°$ while keeping the height and speed constant. The crowd was randomly placed considering the fact that there is no ground truth in real-time.

5. Experimental Results

Blob detection aimed to detect regions, either in a digital image or synthetic image. They were tested on the pre-existing state-of-the-art methods known as: From Open Set to Closed Set: Supervised Spatial Divide-and-Conquer for Object Counting (S-DCNet) [42], Locate, Size and Count: Accurately Resolving People in Dense Crowds via Detection (LSC-CNN) [92], CNN-based Cascaded Multitask Learning of High-level Prior and Density Estimation for Crowd Counting (CMTL) [93], and Single-Image Crowd Counting via Multicolumn Convolutional Neural Network (MCNN) [10]. The estimated count of our data set against the ground truth was promising and presented in the form of MAE and RMSE. Moreover, we demonstrated that the simulator data is compatible and worked appropriately with real-world crowd data.

The simulated images we used demonstrated a high degree of realism and quality that worked with crowd estimation algorithms trained on real images. As demonstrated in Table 5, S-DCNet, MCNN and CMTL showed promising results on our data set against SHA. CMTL performed better and provided the best MAE of 27.6 and RMSE of 34.6.

Table 5. Testing of Our Data against Shanghai Tech Part_A (SHA) using state-of-the-art methods where the highlighted text demonstrates the methods which performed better on our data set.

Methods	SHA		Our Data Set	
	MAE	**RMSE**	**MAE**	**RMSE**
S-DCNet [42]	58.3	95.0	64.4	103.2
LSC-CNN [92]	66.4	117.0	72.7	128.3
CMTL [93]	101.3	152.4	**27.6**	**34.6**
MCNN [10]	110.2	173.2	**57.2**	**72.2**

Comparing the publicly available aerial crowd data sets using individual state-of-the-art methods (Figure 13), our synthetic data set performed comparatively better than the other two data sets (Table 6). A similar number of images were used for testing and chosen randomly. The VisDone2020_CC data performed better than our data set on S-DCNet with a MAE of 71.39 and RMSE of 123.5. However, our data set performed better than the other two data sets in the remaining methods as shown in Table 6 with a lowest MAE of 27.6 and RMSE of 34.6. For an accurate estimation, the original model was trained on a source domain and can be easily transferred to a target domain by fine-tuning only the last two layers of the trained model, which demonstrates good generalisability. To augment the training sample set for training the MCNN, we cropped 16 patches from each image at various locations and each patch was compressed to $\frac{1}{4}$ size of the original image. The pre-training crowd density was very high where it used geometry-adaptive kernels to generate the density maps and calculate the overlapping region density by calculating the average of the generated maps to assist in more accurate estimation.

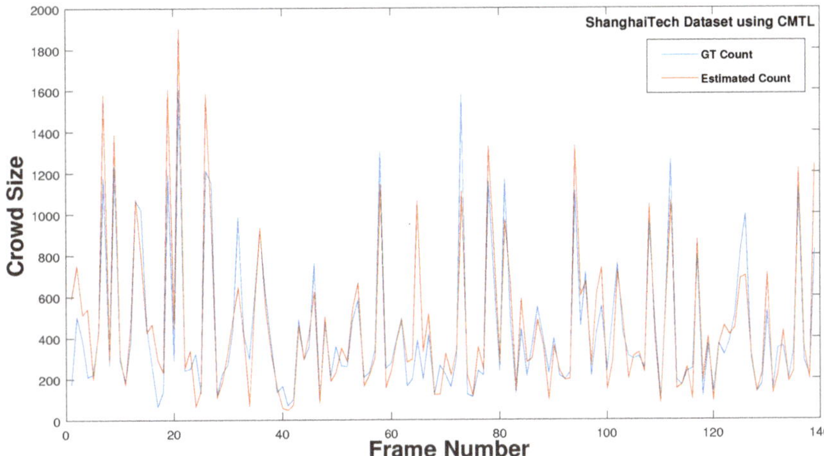

Figure 13. This graph shows the estimated and ground truth count of the CMTL method tested using ShanghaiTech data set.

Table 6. Comparison of aerial crowd data set against state-of-the-art methods.

Methods	DLR_ACD [66]		VisDrone2020-CC [67]		Our Data Set	
	MAE	RMSE	MAE	RMSE	MAE	RMSE
LSC-CNN [92]	71.4	104.3	65.41	107.4	**64.4**	**103.2**
S-DCNet [42]	76.3	134.8	71.39	123.5	72.7	128.3
CMTL [93]	97.2	168.2	103.4	148.2	**27.6**	**34.6**
MCNN [10]	122.1	193.6	118.6	169.6	**57.2**	**72.2**

For any method, data augmentation is important. The S-DCNet results suggest that S-DCNet method is able to adapt to the crowded scenes. The method cropped the original image into 9 sub-images of $\frac{1}{4}$ resolution. Mirroring performance and random scaling doesn't work well on our data. Due to random crowd distribution in our data, the first 4 cropped 224 × 224 sub-images which refers the four corners of the image, didn't fit well and failed to identify the crowded regions in some images which downgraded the performance of our data set. On the other hand, the randomly cropped images improved the downgraded performance and identified the crowded regions which eventually delivered a better performance. However, the VisDrone2020-CC data contains a higher density crowd than ours where the sub-images or cropped patches located the crowd easily. It performed comparatively better on high-density images that justifies that S-DCNet effectively generalises to large crowd data and makes accurate predictions.

After analysing the methods and their best results, we chose CMTL and MCNN for training the model on synthetic data. We selected the CMTL's and MCNN's best model using error on the validation set during training, and set 10% of the training data for validation. Then, we obtained the ground truth density maps using simple Gaussian maps and compared them against network output (Figure 14). The method performed better when the system was trained using synthetic data and tested against the ShanghaiTech data set.

Table 7 shows the output comparison between CMTL model, MCNN model and model trained on our synthetic data set. Our model demonstrates a better performance against the original CMTL model using the same data set which is evident by a low MAE of 98.08 and RMSE of 131.22 which is comparatively better than the original CMTL model. To show the advantage of using our simulator in training with various scenarios, we have additionally trained a multicolumn convolutional neural network (MCNN) on synthetic data and tested against SHA.

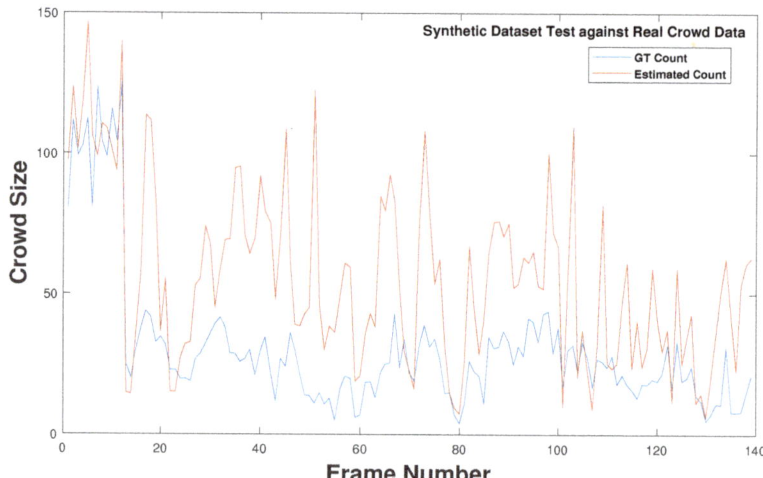

Figure 14. The graph shows the comparison between the ground truth (GT) and the estimated count (ET) that were tested against the CNN-based Cascaded Multitask Learning of High-level Prior and Density Estimation for Crowd Counting (CMTL) [93] method using the aerial synthetic images. We randomly selected 140 images from the synthetic data for testing and compared them against the ground truth.

Table 7. The table presents the results of CMTL model, MCNN model and our synthetic data trained model that were tested against the ShanghaiTech data set.

Method	SHA	
	MAE	**RMSE**
CMTL Model	101.3	152.4
Our Model	**98.08**	**131.22**
MCNN Model	110.2	173.2
Our Model	117.01	194.79

Finally, we tested our own data set as shown in Table 8 using the model trained on synthetic images. The CMTL performed better with the results depicting a lower MAE of 8.58 and RMSE of 10.39. This model offered an accurate estimation of the synthetic data and significantly improved the accuracy of 3D crowd estimation method.

Table 8. The table shows the output of synthetic data model tested against our synthetic data set.

Methods	MAE	RMSE
CMTL	**8.58**	**10.39**
MCNN	17.43	24.46

To the best of our knowledge, this is the first UAV-based system for crowd estimation. The developed system efficiently captures and calculates large crowds spread over a large geographical area. To determine the system's robustness, the results have been compared to standard metrics such as accuracy, recall, RMSE, and MAE. Our proposed method outperforms with a randomly distributed static crowd from a moving camera in 3D and shows a throughput with an accuracy of 89.23%. The output shows the accurate estimation of 116 people out of 130 which highlights the robustness of the proposed method with a possibility to improve the detection rate in further testing. With a precision of 94.30% and recall of 95.86% shown in Table 9, the RMSE of 0.0002748 justifies that the proposed method is efficient to capture and estimate a large geographical area as well as produce an accurate count in minimal time. The method also validated using two-way mapping methods where the output was matched with the ground truth points to cross-check the initial performance.

Table 9. The table shows the results for 3D crowd estimation using UAV method.

RMSE	Accuracy	Precision	Recall	F1
0.0002748	89.23%	94.30%	95.86%	0.9507

Figure 15 illustrates the final output from the ICP [94] where the ground truth points (P) were plotted against the 3D estimated points (MQ). In the ICP, we provided the input point set as P and Q and initialized transform M to be the identity until it was converged. The converged ICP in 1 iteration highlighted the accuracy of the estimated and ground-truth locations. FLANN [91] was used for the nearest neighbour search. Two-way matching of points was carried out and cross-checked between P and MQ. The final result outputs with a list of closest points and 116 pairs matched between P and MQ out of 128 pairs. A wider comparison of our results with the state-of-the-art methods, however, is not possible as no similar method that can justify and motivate us to compare the results with the ground truth exists.

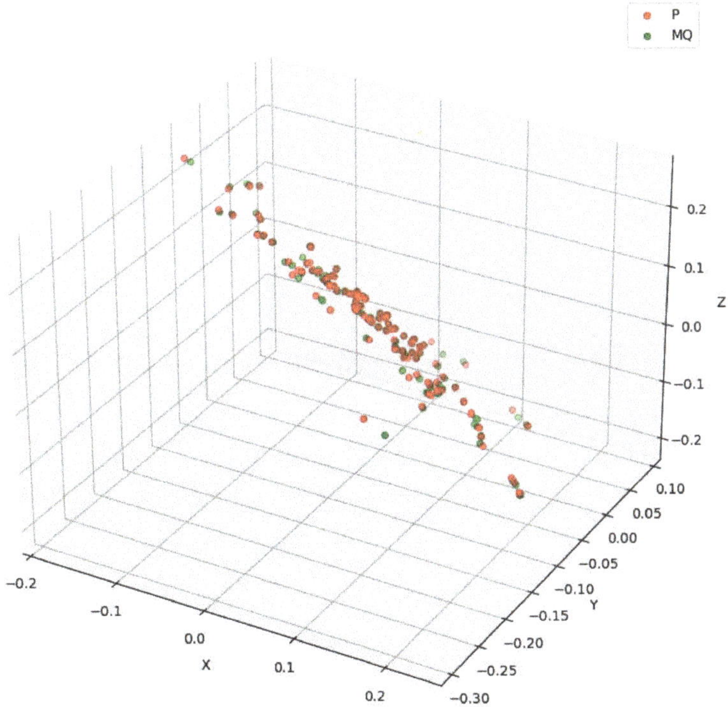

Figure 15. The figure shows the output from the ICP. The plot shows the GT points as P and possible estimated points as MQ.

6. Discussion

The simulation system generated virtual crowd data set was initially tested in conjunction with four well-known state-of-the-art approaches. While performing experiments with virtual crowd data set, we encountered less errors, which is evident by the low MAE and RMSE. The CMTL method outperforms with a MAE of 27.6 and RMSE of 34.6. During the testing, we noted that annotating the accurate position is the most important aspect in accurate computation and generation of density maps. Crowds with distinct features and geometries are important to obtain better results from the virtual data. It not only reduces the chances of overlapping but also helps to create a robust reconstruction model.

In the aerial data set comparison, our simulation system generated data outperforms against DLR_ACD and VisDron2020-CC data sets when tested against S-DCNet. Due to the sparse crowd distribution in our data, the methods which evaluates the entire image as an input such as MCNN and CMTL preforms better than the approaches like S-DCNet that divide the whole image into patches where the accuracy depends on the image density. It should also be noted that the number of patches that lie in the empty region surpasses the crowded region and could not help much in estimation and the error rate will be high.

For a better evaluation of the crowd counting method performance under practical conditions, we have simulated and labeled our new data set. Furthermore, our model has been trained on a source domain that can be easily transferred to a target domain by fine-tuning only the last few layers of the trained model. To enhance the training sample set for training the MCNN, we have cropped 16 patches from each image at various locations and each patch is compressed to $\frac{1}{4}$ size of the original image. Our data set outperforms against the state-of-art CMTL model with a higher throughput and lowest MAE of 98.08 and RMSE of 131.22. We have also tested our data against the model trained on the same set of data which shows a MAE of 8.58 and RMSE of 10.39. This trained model is helpful especially with the same synthetic data and provides a higher accuracy than any other methods but is limited to the same set of data. Further testing needs to be done on the existing publicly available data sets where we want to see how these synthetic data trained model behaves with a new set of data.

The proposed method of 3D crowd estimation system has been tested on various scenes using random crowd distribution. Further testing needs to be done to improve the consistency of the method. Initial test on a moving camera and static crowd provided the accuracy of 89.23% which need to be improve and tested on a large scale. That said, the problem of moving crowd and moving UAV is still being worked on. Here, the reconstruction of a 3D model needs to be considered carefully because the points not aligned properly leads to a false estimation or an output with a lower accuracy. The overlapping of the data and stability of the moving camera is very crucial and needs to be considered while capturing the crowd.

At any given time, the most important issue is optimising the flight path over a wide area to get the most accurate estimate of the available crowd. For example, crowd density may be higher along roads or maybe spilling out radially from the town centre which needs to be dealt in the near future for more accurate estimation. We have captured the data and gathered information for future analysis of different crowd distribution. This data needs to be studied in terms of how synthetic data differs from real data considering domain randomisation, transfer learning and adaptation.

7. Conclusions

Crowd estimation in the 3D domain has grabbed the attention of the computer vision industry, as it provides a more reliable and comprehensive information of the crowd. In this article, we have presented an up-to-date review of open-source simulators and relevant crowd data sets with their shortcomings. It primarily justifies the need of a 3D simulator and explains the type of data the simulator should generate. The paper describes the initial issues of crowd estimation from a moving camera and proposes a solution by developing a 3D crowd simulator for training and testing. It also covers the testing of 3D simulator data by implementing the pre-existing techniques such as LSC-CNN, S-DCNet, CMTL and MCNN. Moreover, it highlights a pre-developed approach to train the synthetic data precisely and validate it using state-of-the-art methods, which justifies that virtual data is as effective as the existing data captured in reality. This will contribute in future development by generating more virtual data sets which could be useful for training deep learning models. In addition, it identifies three big and precise crowd estimation issues, along with introducing a method for 3D crowd estimation using UAV. The presented method can estimate large crowd spread over a large geographical area. Lastly, it explores the limitations that the current model do not address, as well as what needs to be addressed in the future and how the current state will assist in addressing future problems.

In the future, our presented approach could be extended for various potential 3D applications which include tourist attraction [95] using video information to attract and maintain tourist flow, suspicious action detection [96] by monitoring crowded areas and alerting authorities of any suspicious activities and safety monitoring [97] in various facilities, such as religious gatherings, airports, and public areas to monitor crowds, among others.

Author Contributions: Conceptualization, S.S., B.T. and H.C.M.; methodology, S.S. and B.T.; validation, S.S., B.T. and H.C.M.; formal analysis, B.T. and H.C.M.; resources, S.S. and B.T.; data curation, S.S. and B.T.; writing—original draft preparation, S.S.; writing—review and editing, B.T. and H.C.M.; visualization, S.S.; supervision, B.T. and H.C.M. All authors have read and agreed to the published version of the manuscript.

Funding: This research received no external funding.

Institutional Review Board Statement: Not applicable.

Informed Consent Statement: Not applicable.

Data Availability Statement: The data and the details regarding where data supporting reported results in this paper are available from the corresponding author.

Acknowledgments: In this section you can acknowledge any support given which is not covered by the author contribution or funding sections. This may include administrative and technical support, or donations in kind (e.g., materials used for experiments).

Conflicts of Interest: The authors declare no conflict of interest.

Abbreviations

The following abbreviations are used in this manuscript:

2D	2-Dimensional
3D	3-Dimensional
CMTL	Cascaded Multitask Learning
CPU	Central Processing Unit
FLANN	Fast Library for Approximate Nearest Neighbors
GPU	Graphics Processing Unit
GT	Ground Truth
ICP	Iterative Closest Point
LSC-CNN	Locate, Size and Count
MAE	Mean Absolute Error
MCNN	Multicolumn Convolutional Neural Network
MVS	Multiview Stereo
PReLU	Parametric Rectified Linear Unit
RMSE	Root Mean Square Error
S-DCNet	Spatial Divide-and-Conquer
SfM	Structure-from-Motion
UE4	Unreal Engine 4
UAV	Unmanned Aerial Vehicle

References

1. Jacobs, H. To count a crowd. *Columbia J. Rev.* **1967**, *6*, 37.
2. Marsden, M.; McGuinness, K.; Little, S.; O'Connor, N.E. ResnetCrowd: A residual deep learning architecture for crowd counting, violent behaviour detection and crowd density level classification. In Proceedings of the 2017 14th IEEE International Conference on Advanced Video and Signal Based Surveillance (AVSS), Lecce, Italy, 29 August–1 September 2017; pp. 1–7.
3. Loy, C.C.; Chen, K.; Gong, S.; Xiang, T. Crowd counting and profiling: Methodology and evaluation. In *Modeling, Simulation and Visual Analysis of Crowds*; Springer: Berlin/Heidelberg, Germany, 2013; pp. 347–382.
4. Dollar, P.; Wojek, C.; Schiele, B.; Perona, P. Pedestrian detection: An evaluation of the state of the art. *IEEE Trans. Pattern Anal. Mach. Intell.* **2012**, *34*, 743–761. [CrossRef]
5. Li, M.; Zhang, Z.; Huang, K.; Tan, T. Estimating the number of people in crowded scenes by mid based foreground segmentation and head-shoulder detection. In Proceedings of the IEEE 2008 19th International Conference on Pattern Recognition (ICPR), Tampa, FL, USA, 8–11 December 2008; pp. 1–4.
6. Arteta, C.; Lempitsky, V.; Zisserman, A. Counting in the wild. In *European Conference on Computer Vision*; Springer: Cham, Switzerland, 2016; pp. 483–498.
7. Ryan, D.; Denman, S.; Fookes, C.; Sridharan, S. Crowd counting using multiple local features. In Proceedings of the IEEE Digital Image Computing: Techniques and Applications (DICTA'09), Melbourne, Australia, 1–3 December 2009; pp. 81–88.
8. Ma, R.; Li, L.; Huang, W.; Tian, Q. On pixel count based crowd density estimation for visual surveillance. In Proceedings of the 2004 IEEE Conference on Cybernetics and Intelligent Systems, Singapore, 1–3 December 2004; Volume 1, pp. 170–173.
9. Idrees, H.; Soomro, K.; Shah, M. Detecting humans in dense crowds using locally-consistent scale prior and global occlusion reasoning. *IEEE Trans. Pattern Anal. Mach. Intell.* **2015**, *37*, 1986–1998. [CrossRef] [PubMed]
10. Zhang, Y.; Zhou, D.; Chen, S.; Gao, S.; Ma, Y. Single-image crowd counting via multi-column convolutional neural network. In Proceedings of the IEEE Conference on Computer Vision and Pattern Recognition, Las Vegas, NV, USA, 27–30 June 2016; pp. 589–597.
11. Zhang, Q.; Chan, A.B. 3d crowd counting via multi-view fusion with 3d gaussian kernels. In Proceedings of the AAAI Conference on Artificial Intelligence, New York, NY, USA, 7–12 February 2020; Volume 34, pp. 12837–12844. [CrossRef]
12. Zhao, Z.; Shi, M.; Zhao, X.; Li, L. Active Crowd Counting with Limited Supervision. In Proceedings of the Computer Vision-ECCV 2020: 16th European Conference, Glasgow, UK, 23–28 August 2020.
13. Wang, M.; Cai, H.; Han, X.; Zhou, J.; Gong, M. STNet: Scale Tree Network with Multi-level Auxiliator for Crowd Counting. *arXiv* **2020**, arXiv:2012.10189.
14. Ranjan, V.; Wang, B.; Shah, M.; Hoai, M. Uncertainty estimation and sample selection for crowd counting. In Proceedings of the Asian Conference on Computer Vision, Kyoto, Japan, 30 November–4 December 2020.
15. Mustapha, S.; Kassir, A.; Hassoun, K.; Dawy, Z.; Abi-Rached, H. Estimation of crowd flow and load on pedestrian bridges using machine learning with sensor fusion. *Autom. Constr.* **2020**, *112*, 103092. [CrossRef]
16. Almeida, I.; Jung, C. Crowd flow estimation from calibrated cameras. *Mach. Vis. Appl.* **2021**, *32*, 1–12. [CrossRef]
17. Choi, H.; Moon, G.; Park, J.; Lee, K.M. 3DCrowdNet: 2D Human Pose-Guided3D Crowd Human Pose and Shape Estimation in the Wild. *arXiv* **2021**, arXiv:2104.07300.
18. Fahad, M.S.; Deepak, A. Crowd Estimation of Real-Life Images with Different View-Points. In Proceedings of the International Conference on Innovative Computing and Communications, Delhi, India, 21–23 February 2021; pp. 1053–1062.
19. Zhan, B.; Monekosso, D.N.; Remagnino, P.; Velastin, S.A.; Xu, L.Q. Crowd analysis: A survey. *Mach. Vis. Appl.* **2008**, *19*, 345–357. [CrossRef]
20. Junior, J.C.S.J.; Musse, S.R.; Jung, C.R. Crowd analysis using computer vision techniques. *IEEE Signal Process. Mag.* **2010**, *27*, 66–77.
21. Teixeira, T.; Dublon, G.; Savvides, A. A survey of human-sensing: Methods for detecting presence, count, location, track, and identity. *ACM Comput. Surv.* **2010**, *5*, 59–69.
22. Ferryman, J.; Ellis, A.L. Performance evaluation of crowd image analysis using the PETS2009 dataset. *Pattern Recognit. Lett.* **2014**, *44*, 3–15. [CrossRef]
23. Li, T.; Chang, H.; Wang, M.; Ni, B.; Hong, R.; Yan, S. Crowded scene analysis: A survey. *IEEE Trans. Circuits Syst. Video Technol.* **2014**, *25*, 367–386. [CrossRef]
24. Hu, W.; Tan, T.; Wang, L.; Maybank, S. A survey on visual surveillance of object motion and behaviors. *IEEE Trans. Syst. Man Cybern. Part C Appl. Rev.* **2004**, *34*, 334–352. [CrossRef]
25. Ryan, D.; Denman, S.; Sridharan, S.; Fookes, C. An evaluation of crowd counting methods, features and regression models. *Comput. Vis. Image Underst.* **2015**, *130*, 1–17. [CrossRef]
26. Chan, A.B.; Liang, Z.S.J.; Vasconcelos, N. Privacy preserving crowd monitoring: Counting people without people models or tracking. In Proceedings of the 2008 IEEE Conference on Computer Vision and Pattern Recognition (CVPR), Anchorage, AK, USA, 23–28 June 2008; pp. 1–7.
27. Ferryman, J.; Shahrokni, A. Pets2009: Dataset and challenge. In Proceedings of the 2009 Twelfth IEEE International Workshop on Performance Evaluation of Tracking and Surveillance, Snowbird, UT, USA, 7–9 December 2009; pp. 1–6.
28. Tan, B.; Zhang, J.; Wang, L. Semi-supervised elastic net for pedestrian counting. *Pattern Recognit.* **2011**, *44*, 2297–2304. [CrossRef]
29. Chen, K.; Loy, C.C.; Gong, S.; Xiang, T. Feature mining for localised crowd counting. *BMVC* **2012**, *1*, 3.

30. Zhou, B.; Wang, X.; Tang, X. Understanding collective crowd behaviors: Learning a mixture model of dynamic pedestrian-agents. In Proceedings of the 2012 IEEE Conference on Computer Vision and Pattern Recognition, Providence, RI, USA, 16–21 June 2012; pp. 2871–2878.
31. Saleh, S.A.M.; Suandi, S.A.; Ibrahim, H. Recent survey on crowd density estimation and counting for visual surveillance. *Eng. Appl. Artif. Intell.* **2015**, *41*, 103–114. [CrossRef]
32. Zitouni, M.S.; Bhaskar, H.; Dias, J.; Al-Mualla, M.E. Advances and trends in visual crowd analysis: A systematic survey and evaluation of crowd modelling techniques. *Neurocomputing* **2016**, *186*, 139–159. [CrossRef]
33. Grant, J.M.; Flynn, P.J. Crowd scene understanding from video: A survey. *ACM Trans. Multimed. Comput. Commun. Appl. (TOMM)* **2017**, *13*, 1–23. [CrossRef]
34. Sindagi, V.A.; Patel, V.M. A survey of recent advances in cnn-based single image crowd counting and density estimation. *Pattern Recognit. Lett.* **2018**, *107*, 3–16. [CrossRef]
35. Walach, E.; Wolf, L. Learning to count with cnn boosting. In Proceedings of the European Conference on Computer Vision, Amsterdam, The Netherlands, 11–14 October 2016; Springer: Berlin/Heidelberg, Germany, 2016; pp. 660–676.
36. Shang, C.; Ai, H.; Bai, B. End-to-end crowd counting via joint learning local and global count. In Proceedings of the 2016 IEEE International Conference on Image Processing (ICIP), Phoenix, AZ, USA, 25–28 September 2016; pp. 1215–1219.
37. Onoro-Rubio, D.; López-Sastre, R.J. Towards perspective-free object counting with deep learning. In Proceedings of the European Conference on Computer Vision, Amsterdam, The Netherlands, 11–14 October 2016; Springer: Berlin/Heidelberg, Germany, 2016; pp. 615–629.
38. Kang, D.; Ma, Z.; Chan, A.B. Beyond counting: Comparisons of density maps for crowd analysis tasks—Counting, detection, and tracking. *IEEE Trans. Circuits Syst. Video Technol.* **2018**, *29*, 1408–1422. [CrossRef]
39. Tripathi, G.; Singh, K.; Vishwakarma, D.K. Convolutional neural networks for crowd behaviour analysis: A survey. *Vis. Comput.* **2019**, *35*, 753–776. [CrossRef]
40. Gao, G.; Gao, J.; Liu, Q.; Wang, Q.; Wang, Y. Cnn-based density estimation and crowd counting: A survey. *arXiv* **2020**, arXiv:2003.12783.
41. Yan, Z.; Yuan, Y.; Zuo, W.; Tan, X.; Wang, Y.; Wen, S.; Ding, E. Perspective-guided convolution networks for crowd counting. In Proceedings of the IEEE/CVF International Conference on Computer Vision, Seoul, Korea, 27 October–2 November 2019; pp. 952–961.
42. Xiong, H.; Lu, H.; Liu, C.; Liang, L.; Cao, Z.; Shen, C. From Open Set to Closed Set: Counting Objects by Spatial Divide-and-Conquer. In Proceedings of the IEEE/CVF International Conference on Computer Vision (ICCV), Seoul, Korea, 27 October–2 November 2019; pp. 8362–8371.
43. Tian, Y.; Lei, Y.; Zhang, J.; Wang, J.Z. Padnet: Pan-density crowd counting. *IEEE Trans. Image Process.* **2019**, *29*, 2714–2727. [CrossRef] [PubMed]
44. Kleinmeier, B.; Zönnchen, B.; Gödel, M.; Köster, G. Vadere: An open-source simulation framework to promote interdisciplinary understanding. *arXiv* **2019**, arXiv:1907.09520.
45. Maury, B.; Faure, S. *Crowds in Equations: An Introduction to the Microscopic Modeling of Crowds*; World Scientific: Singapore, 2018.
46. Curtis, S.; Best, A.; Manocha, D. Menge: A modular framework for simulating crowd movement. *Collect. Dyn.* **2016**, *1*, 1–40. [CrossRef]
47. Wagoum, A.K.; Chraibi, M.; Zhang, J.; Lämmel, G. JuPedSim: An open framework for simulating and analyzing the dynamics of pedestrians. In Proceedings of the 3rd Conference of Transportation Research Group of India, Kolkata, India, 17–20 December 2015; Volume 12.
48. Grimm, V.; Revilla, E.; Berger, U.; Jeltsch, F.; Mooij, W.M.; Railsback, S.F.; Thulke, H.H.; Weiner, J.; Wiegand, T.; DeAngelis, D.L. Pattern-oriented modeling of agent-based complex systems: Lessons from ecology. *Science* **2005**, *310*, 987–991. [CrossRef]
49. Van Den Berg, J.; Patil, S.; Sewall, J.; Manocha, D.; Lin, M. Interactive navigation of multiple agents in crowded environments. In Proceedings of the 2008 Symposium on Interactive 3D Graphics and Games, Redwood City, CA, USA, 15–17 February 2008; pp. 139–147.
50. McGrattan, K.; Hostikka, S.; McDermott, R.; Floyd, J.; Weinschenk, C.; Overholt, K. Fire dynamics simulator user's guide. *NIST Spec. Publ.* **2013**, *1019*, 1–339.
51. Idrees, H.; Saleemi, I.; Seibert, C.; Shah, M. Multi-source multi-scale counting in extremely dense crowd images. In Proceedings of the IEEE Conference on Computer Vision and Pattern Recognition, Portland, OR, USA, 23–28 June 2013; pp. 2547–2554.
52. Sindagi, V.A.; Patel, V.M. Generating high-quality crowd density maps using contextual pyramid cnns. In Proceedings of the 2017 IEEE International Conference on Computer Vision (ICCV), Venice, Italy, 22–29 October 2017; pp. 1879–1888.
53. Wang, Q.; Gao, J.; Lin, W.; Li, X. NWPU-Crowd: A Large-Scale Benchmark for Crowd Counting and Localization. *IEEE Trans. Pattern Anal. Mach. Intell.* **2020**. [CrossRef]
54. Sindagi, V.A.; Yasarla, R.; Patel, V.M. JHU-CROWD++: Large-Scale Crowd Counting Dataset and A Benchmark Method. *IEEE Trans. Pattern Anal. Mach. Intell.* **2020**. [CrossRef]
55. Idrees, H.; Tayyab, M.; Athrey, K.; Zhang, D.; Al-Maadeed, S.; Rajpoot, N.; Shah, M. Composition loss for counting, density map estimation and localization in dense crowds. In Proceedings of the European Conference on Computer Vision (ECCV), Glasgow, UK, 23–28 August 2018; pp. 532–546.

56. Hu, D.; Mou, L.; Wang, Q.; Gao, J.; Hua, Y.; Dou, D.; Zhu, X.X. Ambient Sound Helps: Audiovisual Crowd Counting in Extreme Conditions. *arXiv* **2020**, arXiv:2005.07097.

57. Lian, D.; Li, J.; Zheng, J.; Luo, W.; Gao, S. Density Map Regression Guided Detection Network for RGB-D Crowd Counting and Localization. In Proceedings of the IEEE Conference on Computer Vision and Pattern Recognition (CVPR), Long Beach, CA, USA, 15–20 June 2019.

58. Fang, Y.; Zhan, B.; Cai, W.; Gao, S.; Hu, B. Locality-constrained Spatial Transformer Network for Video Crowd Counting. *arXiv* **2019**, arXiv:1907.07911.

59. Wang, Q.; Gao, J.; Lin, W.; Yuan, Y. Learning from Synthetic Data for Crowd Counting in the Wild. In Proceedings of the IEEE Conference on Computer Vision and Pattern Recognition (CVPR), Long Beach, CA, USA, 16–20 June 2019; pp. 8198–8207.

60. Liu, W.; Lis, K.M.; Salzmann, M.; Fua, P. Geometric and Physical Constraints for Drone-Based Head Plane Crowd Density Estimation. In Proceedings of the IEEE/RSJ International Conference on Intelligent Robots and Systems (IROS), Macau, China, 3–8 November 2019.

61. Zhang, Q.; Chan, A.B. Wide-Area Crowd Counting via Ground-Plane Density Maps and Multi-View Fusion CNNs. In Proceedings of the IEEE Conference on Computer Vision and Pattern Recognition, Long Beach, CA, USA, 15–20 June 2019; pp. 8297–8306.

62. Deng, L.; Wang, S.H.; Zhang, Y.D. Fully Optimized Convolutional Neural Network Based on Small-Scale Crowd. In Proceedings of the 2020 IEEE International Symposium on Circuits and Systems (ISCAS), Seville, Spain, 12–14 October 2020; pp. 1–5.

63. Mallapuram, S.; Ngwum, N.; Yuan, F.; Lu, C.; Yu, W. Smart city: The state of the art, datasets, and evaluation platforms. In Proceedings of the 2017 IEEE/ACIS 16th International Conference on Computer and Information Science (ICIS), Wuhan, China, 24–26 May 2017; pp. 447–452.

64. Lim, M.K.; Kok, V.J.; Loy, C.C.; Chan, C.S. Crowd saliency detection via global similarity structure. In Proceedings of the IEEE 2014 22nd International Conference on Pattern Recognition, Stockholm, Sweden, 24–28 August 2014; pp. 3957–3962.

65. Zhang, C.; Kang, K.; Li, H.; Wang, X.; Xie, R.; Yang, X. Data-driven crowd understanding: A baseline for a large-scale crowd dataset. *IEEE Trans. Multimed.* **2016**, *18*, 1048–1061. [CrossRef]

66. Bahmanyar, R.; Vig, E.; Reinartz, P. MRCNet: Crowd counting and density map estimation in aerial and ground imagery. *arXiv* **2019**, arXiv:1909.12743.

67. Zhu, P.; Wen, L.; Du, D.; Bian, X.; Hu, Q.; Ling, H. Vision Meets Drones: Past, Present and Future. *arXiv* **2020**, arXiv:2001.06303.

68. Zhu, P.; Sun, Y.; Wen, L.; Feng, Y.; Hu, Q. Drone Based RGBT Vehicle Detection and Counting: A Challenge. *arXiv* **2020**, arXiv:2003.02437.

69. Chen, L.; Wang, G.; Hou, G. Multi-scale and multi-column convolutional neural network for crowd density estimation. *Multimed. Tools Appl.* **2021**, *80*, 6661–6674. [CrossRef]

70. Guo, L.; Zhou, W. Crowd Density Estimation Based on Multi-Column Hybrid Convolutional Network. *J. Phys. Conf. Ser.* **2021**, *1828*, 012025. [CrossRef]

71. Jingying, W. A Survey on Crowd Counting Methods and Datasets. In *Advances in Computer, Communication and Computational Sciences*; Springer: Berlin/Heidelberg, Germany, 2021; pp. 851–863.

72. Ma, Y.J.; Shuai, H.H.; Cheng, W.H. Spatiotemporal Dilated Convolution with Uncertain Matching for Video-based Crowd Estimation. *IEEE Trans. Multimed.* **2021**. [CrossRef]

73. Chen, H.; Guo, P.; Li, P.; Lee, G.H.; Chirikjian, G. Multi-person 3D Pose Estimation in Crowded Scenes Based on Multi-view Geometry. In Proceedings of the European Conference on Computer Vision, Glasgow, UK, 23–28 August 2020; Springer: Berlin/Heidelberg, Germany, 2020; pp. 541–557.

74. Benzine, A.; Chabot, F.; Luvison, B.; Pham, Q.C.; Achard, C. Pandanet: Anchor-based single-shot multi-person 3d pose estimation. In Proceedings of the IEEE/CVF Conference on Computer Vision and Pattern Recognition, Seattle, WA, USA, 13–19 June 2020; pp. 6856–6865.

75. Song, J.Y.; Chung, J.J.Y.; Fouhey, D.F.; Lasecki, W.S. C-Reference: Improving 2D to 3D Object Pose Estimation Accuracy via Crowdsourced Joint Object Estimation. *Proc. ACM Hum.-Comput. Interact.* **2020**, *4*, 1–28. [CrossRef]

76. Chen, H.; Guo11, P.; Li, P.; Lee, G.H.; Chirikjian, G. Multi-person 3D Pose Estimation in Crowded Scenes Based on Multi-View Geometry—Supplementary Material. Available online: https://www.ecva.net/papers/eccv_2020/papers_ECCV/papers/1234 80545.pdf (accessed on 10 January 2021).

77. Hashmi, M.F.; Ashish, B.K.K.; Keskar, A.G. GAIT analysis: 3D pose estimation and prediction in defence applications using pattern recognition. In Proceedings of the Twelfth International Conference on Machine Vision (ICMV 2019), International Society for Optics and Photonics, Amsterdam, The Netherlands, 16–18 November 2020; Volume 11433, p. 114330S.

78. Zheng, C.; Zhu, S.; Mendieta, M.; Yang, T.; Chen, C.; Ding, Z. 3d human pose estimation with spatial and temporal transformers. *arXiv* **2021**, arXiv:2103.10455.

79. Li, W.; Liu, H.; Ding, R.; Liu, M.; Wang, P. Lifting Transformer for 3D Human Pose Estimation in Video. *arXiv* **2021**, arXiv:2103.14304.

80. Kumarapu, L.; Mukherjee, P. Animepose: Multi-person 3d pose estimation and animation. *Pattern Recognit. Lett.* **2021**, *147*, 16–24. [CrossRef]

81. Epic Games; Unreal Engine/The Most Powerful Real-Time 3D Creation Platform. Available online: https://www.unrealengine.com (accessed on 10 January 2021).

82. Human, M. Make Human. Available online: http://www.makehumancommunity.org/ (accessed on 22 February 2021).

83. AXYZ; Anima. Available online: https://secure.axyz-design.com/ (accessed on 5 March 2021).

84. Schönberger, J.L.; Frahm, J.M. Colmap/Structure-from-Motion Revisited. Available online: https://colmap.github.io/ (accessed on 21 April 2021).

85. Schops, T.; Schonberger, J.L.; Galliani, S.; Sattler, T.; Schindler, K.; Pollefeys, M.; Geiger, A. A multi-view stereo benchmark with high-resolution images and multi-camera videos. In Proceedings of the IEEE Conference on Computer Vision and Pattern Recognition, Honolulu, HI, USA, 21–26 July 2017; pp. 3260–3269.

86. Stathopoulou, E.K.; Remondino, F. Open-source image-based 3D reconstruction pipelines: Review, comparison and evaluation. In Proceedings of the 6th International Workshop LowCost 3D—Sensors, Algorithms, Applications, Strasbourg, France, 2–3 December 2019; ISPRS: Strasbourg, France, 2019; pp. 331–338.

87. Leudet, J.; Mikkonen, T.; Christophe, F.; Männistö, T. Virtual Environment for Training Autonomous Vehicles. In Proceedings of the Annual Conference Towards Autonomous Robotic Systems, Bristol, UK, 25–27 July 2018; Springer: Berlin/Heidelberg, Germany, 2018; pp. 159–169.

88. Kazhdan, M.; Hoppe, H. Screened poisson surface reconstruction. *ACM Trans. Graph. (ToG)* **2013**, *32*, 1–13. [CrossRef]

89. Contributor, Wiki. Iterative Closest Point. Available online: https://en.wikipedia.org/wiki/Iterative_Closest_Point (accessed on 24 June 2021).

90. Marden, S.; Guivant, J. Improving the performance of ICP for real-time applications using an approximate nearest neighbour search. In Proceedings of Australasian Conference on Robotics and Automation, Wellington, New Zealand, 3–5 December 2012; pp. 1–6.

91. Muja, M.; Lowe, D.G. Fast approximate nearest neighbors with automatic algorithm configuration. *VISAPP (1)* **2009**, *2*, 2.

92. Sam, D.B.; Peri, S.V.; Sundararaman, M.N.; Kamath, A.; Radhakrishnan, V.B. Locate, Size and Count: Accurately Resolving People in Dense Crowds via Detection. *IEEE Trans. Pattern Anal. Mach. Intell.* **2020**. [CrossRef]

93. Sindagi, V.A.; Patel, V.M. Cnn-based cascaded multi-task learning of high-level prior and density estimation for crowd counting. In Proceedings of the 2017 14th IEEE International Conference on Advanced Video and Signal Based Surveillance (AVSS), Lecce, Italy, 29 August–1 September 2017; pp. 1–6.

94. Liu, H.; Wang, S.; Zhao, D. Initial alignment for point cloud registration by improved differential evolution algorithm. *Optik* **2021**, *243*, 166856. [CrossRef]

95. Li, L. A Crowd Density Detection Algorithm for Tourist Attractions Based on Monitoring Video Dynamic Information Analysis. *Complexity* **2020**, *2020*, 6635446. [CrossRef]

96. Penmetsa, S.; Minhuj, F.; Singh, A.; Omkar, S. Autonomous UAV for suspicious action detection using pictorial human pose estimation and classification. *ELCVIA Electron. Lett. Comput. Vis. Image Anal.* **2014**, *13*, 0018–0032. [CrossRef]

97. Zhou, B.; Tang, X.; Wang, X. Learning collective crowd behaviors with dynamic pedestrian-agents. *Int. J. Comput. Vis.* **2015**, *111*, 50–68. [CrossRef]

Article

UAV-Assisted Wide Area Multi-Camera Space Alignment Based on Spatiotemporal Feature Map

Jing Li [1,*], Yuguang Xie [1], Congcong Li [1], Yanran Dai [1], Jiaxin Ma [1], Zheng Dong [2] and Tao Yang [2]

[1] School of Telecommunications Engineering, Xidian University, Xi'an 710071, China; ygxie@stu.xidian.edu.cn (Y.X.); ccli@stu.xidian.edu.cn (C.L.); yrdai@stu.xidian.edu.cn (Y.D.); jxma_0@stu.xidian.edu.cn (J.M.)
[2] National Engineering Laboratory for Integrated Aero-Space-Ground-Ocean Big Data Application Technology, SAIIP School of Computer Science, Northwestern Polytechnical University, Xi'an 710129, China; dongzheng@mail.nwpu.edu.cn (Z.D.); tyang@nwpu.edu.cn (T.Y.)
* Correspondence: jinglixd@mail.xidian.edu.cn; Tel.: +86-139-9132-0168

Abstract: In this paper, we investigate the problem of aligning multiple deployed camera into one united coordinate system for cross-camera information sharing and intercommunication. However, the difficulty is greatly increased when faced with large-scale scene under chaotic camera deployment. To address this problem, we propose a UAV-assisted wide area multi-camera space alignment approach based on spatiotemporal feature map. It employs the great global perception of Unmanned Aerial Vehicles (UAVs) to meet the challenge from wide-range environment. Concretely, we first present a novel spatiotemporal feature map construction approach to represent the input aerial and ground monitoring data. In this way, the motion consistency across view is well mined to overcome the great perspective gap between the UAV and ground cameras. To obtain the corresponding relationship between their pixels, we propose a cross-view spatiotemporal matching strategy. Through solving relative relationship with the above air-to-ground point correspondences, all ground cameras can be aligned into one surveillance space. The proposed approach was evaluated in both simulation and real environments qualitatively and quantitatively. Extensive experimental results demonstrate that our system can successfully align all ground cameras with very small pixel error. Additionally, the comparisons with other works on different test situations also verify its superior performance.

Keywords: multi-camera system; space alignment; UAV-assisted calibration; cross-view matching; spatiotemporal feature map; view-invariant description; air-to-ground synchronization

Citation: Li, J.; Xie, Y.; Li, C.; Dai, Y.; Ma, J.; Dong, Z.; Yang, T. UAV-Assisted Wide Area Multi-Camera Space Alignment Based on Spatiotemporal Feature Map. *Remote Sens.* **2021**, *13*, 1117. https://doi.org/10.3390/rs13061117

Academic Editor: Anwaar Ulhaq

Received: 2 February 2021
Accepted: 11 March 2021
Published: 15 March 2021

1. Introduction

The advance of imaging performance and decline of sensor price play a significant role in promoting the popularization and development of multi-camera systems. With its advantages, such as complementary field of view, flexible structural arrangement and diverse acquisition forms, multi-camera systems have an increasingly important effect in the field of security surveillance [1,2], automatic controlling [3,4], intelligent transportation [5,6], etc. Among them, camera space alignment, which is the foundation and difficulty for large-scale multi-camera systems, has gradually become one of the research focuses in recent years. It aims to unify visual data from different cameras into one coordinate system which contributes to cross-camera information sharing and interconnection.

To date, several related algorithms have put been forward for camera spatial relationship estimation of multi-camera system space alignment [7–9]. According to whether the camera field of view overlaps, numerous corresponding space alignment solutions are presented for overlapping cameras and non-overlapping cameras, respectively. When there are overlapping areas between cameras, we can use common features from additional calibrator or only own scene to calculate the relative camera relationship matrix for space alignment. There are many and various types of calibration object: one-dimensional calibrating bar, board calibration plane, stereo calibration tower, etc. For space alignment of cameras without overlapping, current approaches relate these independent but closely linked visual data by intermediate connector, e.g., scene 3D map, mirror reflection, moving target, common marker, etc. Their performances typically rely on the accuracy and robustness of cross-camera link bridge establishment. Based on the above achievements, several technical issues such as active tracking and situation awareness can be studied and implemented under multi-camera spatial calibration results.

However, despite recent advances, there are still many problems that need further research on existing deployed multi-camera space alignment. The main difficulties cover the following points: (1) Chaotic spatial layout: Most cameras are set up at different times for different application requirements. Lack of scientific topology structure planning and design lead to chaotic layout. Thus, the overlapping relation between cameras is also complex. (2) Large scale environment: Multi-camera systems are mostly used in large scenes because of their wider coverage. Thus, specially designed calibrators with limited size and fixed shape are inapplicable. Meanwhile, how to balance accuracy and efficiency in large-scale environment is also a challenge. (3) Great visual gap: Cameras are distributed dispersedly under wide baseline. There are differences between cameras in viewing angle, rotation and object scale. These differences bring great difficulty on space alignment across cameras.

In this paper, we thoroughly analyze the above problems of multi-camera space alignment in large-scale environment. Its essence lies in how to better build the connection among these independent cameras. This problem, in a sense, is similar to multi-station cooperative wireless communication [10]. In its relevant studies, a UAV is employed as relay node to maintain stable signal coverage in long-distance data transmission due to its mobility and flexibility. Inspired by this, we extend the thinking of UAV assistance to multi-camera space alignment, as shown in Figure 1. However, UAV airborne camera and ground deployed camera observe the surveillance scene in aerial view and street view, respectively. Significant perspective differences make it hard to directly match the air with the ground. To address this problem, we explore the consistency of motion across different views. Based on the principle that intersection point is invariable under projection transformation, we construct spatiotemporal feature map which records the time and position of intersection generated by moving targets. Through matching these feature maps, time synchronization and spatial alignment can be achieved simultaneously. The relative relationship between ground cameras and UAV is established. Multiple cameras are aligned into one coordination system with the auxiliary connection of UAV.

Following the above research route, we propose a novel UAV-assisted multi-camera space alignment algorithm based on spatiotemporal feature map. Concretely, it contains two main modules: one is spatiotemporal feature map construction to describe UAV-assisted aerial data and ground monitoring data and the other is cross-view spatiotemporal matching based on feature map. The first one employs several lines perpendicular to the road direction as the feature detection lines. The corresponding spatiotemporal feature map can be constructed by recording the time and position of moving target crossing each line. On this basis, we then present a novel cross-view matching strategy which deeply explores their relations through the waveform change of time series and space distribution. With UAV-to-ground matching point pairs, we can calibrate ground cameras' space relationship to UAV. When the spatial parameters of all ground cameras are estimated, the multi-camera system is aligned into one united space under UAV assistance.

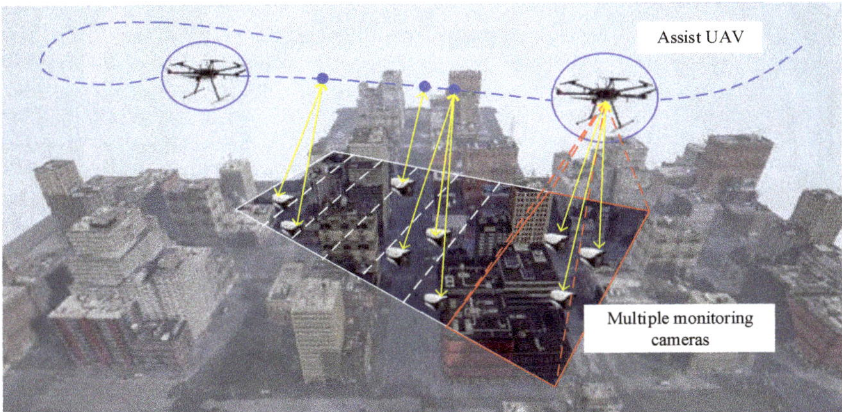

Figure 1. An illustration of UAV-assisted wide area multi-camera space alignment. Through air-to-ground matching based on spatiotemporal feature map, the relative relationship between UAV and ground cameras is obtained (yellow line). Since we can unity the UAV's external parameters (blue line), multiple cameras in a large-scale environment are aligned into one coordination system with UAV auxiliary linkage effectively and efficiently.

1.1. Related Work

In this section, we review the other multi-camera space calibration works which are related to the proposed method. Multi-camera space alignment calibrates all sensors together by estimating each sensor's rotation matrix and translation matrix in one reference coordinate system. According to whether there is overlap between their field of view, we divide existing methods into two categories: overlapping cameras and non-overlapping cameras.

For the first one, most scholars mine the common and independent visual data captured by different cameras to estimate their spatial relationship. Either the scene itself or additional calibrator can be used. Many studies are conducted based on the common visual feature of observation scene itself. For example, Lv et al. [11] detected moving humans, who represent common visual information across cameras, and regarded them as a set of sticks with the same height for camera calibration based on vanishing point theory. Liu et al. [12,13] put forward an automatic camera calibration approach and its improvement method using common pedestrian feature. Their methods are proposed under the assumption that all humans are on one plane surface. Unlike them, Truong et al. [7] employed president tracks to match corresponding information in partial overlapped cameras and then computed the extrinsic calibration matrices. Besides these methods using pedestrian information, Romil et al. [14] analyzed the traffic scenarios and introduced a novel camera calibration method by leveraging vehicle feature correspondences between real size and pixel distance. Furthermore, many studies focus on adding common visual information by additional calibration markers [15]: one-dimensional calibration bar, checkerboard plane, stereo calibration tower, etc. One of the widely used calibration algorithms was proposed by Zhang [16], who used single checkerboard calibration plane to estimate camera external and internal parameters simultaneously. Based on Zhang's approach, many corresponding improved methods [17,18] are presented to optimize different parts such as optimization function and calibration object. To overcome the limited stereo information of 2D calibration object, 3D marker is used to camera imaging parameter estimation. Andreas et al. [19] calculated the extrinsic matrix of a multi-camera system with 3D target and then optimized these parameters based on genetic algorithm. Huang et al. [20] designed a cube calibration object which can easily be captured by multiple cameras, and this approach calibrates all cameras in one process with high efficiency and convenience. In summary, the calibration methods of overlapping cameras, whether based on its own scene feature or additional calibrator, have their own advantages and disadvantages. The approach based on the scene feature itself is strongly influenced by the accuracy of feature detection and matching, while the approach based on additional marker usually has poor universality.

Calibration algorithms of non-overlapping cameras can be broadly classified into the following kinds: SLAM-based method, mirror-based method, tracking-based method and marker-based method. Taking advantages of SLAM in visual localization, a user can estimate camera relative pose by several corresponding points. For instance, Yin et al. [21] constructed 3D feature point map of the natural environment. The extrinsic matrix is obtained through the 3D scene point map created by SLAM. Feng et al. [22] modeled the surveillance space by SLAM previously and then employed 2D–3D matching to calibrate camera external parameters. Another extensively applied calibration strategy is based on specular reflection. It can generate the common view between different cameras by planer mirror. Xu et al. [23] employed mirrored phase target as an intermediate linkage, and camera calibration without overlapping can be achieved through mirror reflection relationship. By combining camera projection model and flat refractive geometry, an accurate multiple camera pose estimation approach [24] is investigated with a transparent glass calibration board. Beyond that, some works connect non-overlapping camera with moving object. Sarmadi et al. [25] analyzed the interaction relationship between camera pose estimation and object tracking. Their method shows accurate results on camera imaging parameters estimation and real-time tracking with low computational cost. Similar to overlapping camera calibration, users can also add an extra calibrator. Izaak et al. [26] established a gray code and projected it into a plane with a projector. They could calculate the relative pose between camera, plane and projector. For non-overlapping cameras in aero photogrammetry, Yin et al. [27] introduced a novel marker-based method based on multiple chessboard targets. Sufficient equations can be obtained to solve the extrinsic parameters by moving camera at multiple positions. Recently, Jeong et al. [28] regarded road markings as robust visual feature in urban environment. They realized calibration through joint optimization of normalized information distance, edge alignment and plane fitting. Overall, these algorithms start from different perspectives to solve various problems when calibrating the camera without field of view overlapping.

1.2. Main Contribution

This paper aims to align all deployed monitoring cameras into a united coordinate system. Compared to the aforementioned related works, there are some differences between our proposed approach and them. The problem studied in this paper is more complicated due to the chaotic layout of deployed cameras. The overlapping relationship between cameras is unknown. Meanwhile, the research ideas are also different. Most current strategies employ designed calibrators or scene visual feature to relate multiple cameras, while this paper utilizes UAV as an aid. We give full play to the UAV's global perception ability to cover the challenge in large scenes. In addition, unlike the above methods based on visual features (texture, object trajectory, etc.), we explore a more stable cross-view feature description method based on motion intersection invariance to overcome perspective gap between aerial and ground data. In this paper, we start our research from a new angle and propose a novel UAV-assisted wide area multi-camera space alignment approach.

We summarize our contributions in this paper as follows:

- We propose a multi-camera wide-area space alignment approach with UAV assistance to realize the unification of cameras' imaging coordinate system. Unlike current additional marker-based methods, this paper employs UAV to build visual connection across cameras which shows superior flexibility and efficiency in large-scale environment.

- We present a novel cross-view feature description algorithm, called spatiotemporal feature map, to overcome perspective gap between aerial-view images captured by UAV and street-view images collected by ground cameras. It makes full use of motion consistency among different views, which can implement synchronization on both time and space.

- To better evaluate the proposed method, we establish a new traffic monitoring database collected in both simulation and real environment. This database provides abundant monitoring data captured by multiple cameras at different fixed positions from various scenarios, including crossroad, T-junction, straight road, multi-lane road, etc. Extensive experiments demonstrate that our system returns encouraging space alignment results.

The rest manuscript is organized as follows. A detailed introduction of the proposed approach is described in Section 2. Section 3 evaluates our method in simulation and real-world environment qualitatively and quantitatively. In addition, we also conduct contrast experiments with other methods for performance comparison in Section 4. The parameter influence of system performance is discussed at the end of this section. Finally, Section 5 concludes this paper considering the methodology and experimental results.

2. UAV-Assisted Wide Area Multi-Camera Space Alignment Based on Spatiotemporal Feature Map

Figure 2 provides an overview of the proposed UAV-assisted wide area multi-camera space alignment approach intuitively. With the videos from assisted-UAV and ground monitoring cameras as input, we first describe them by the spatiotemporal feature map, which lays a basis for multi-camera space alignment. Then, this paper puts froward a cross-view spatiotemporal matching strategy to mine the association relationship between these feature maps from multiple levels. The corresponding pixels between UAV-assisted videos and ground fixed videos can be obtained, and then multiple ground cameras are aligned into one surveillance space under UAV auxiliary data connection.

The following notations are used in this manuscript (Table 1).

Table 1. Major notations.

Notation	Description
N	The number of ground monitoring cameras
M	The number of UAV's hovering positions
$V_{C1}, V_{C2}, ... V_{CN}$	The set of ground monitoring videos
$V_{A1}, V_{A2}, ... V_{AM}$	The set of UAV assisted videos
V	An example of monitoring video
N_i	The number of frames obtained from V deframing
NL_i	The number of feature lines detected from V
fl_i	An example of feature line in V
Ng	The number of ground spatiotemporal feature maps
Fg	The set of ground spatiotemporal feature maps
Fg_i	ith ground feature map
Fa_k	kth aerial feature map
$\mathbf{fg}$	The set of feature vectors of Fg in time dimension
$\mathbf{fg}_{ti}$	Feature vector of Fg_i in time dimension
$\mathbf{fa}_{tk}$	Feature vector of Fa_k in time dimension
τ	Time delay
Fg_j'	ith ground feature map after cutting
Fa_k'	kth aerial feature map after cutting
$\mathbf{fg}_{si}$	Feature vector of Fg_j' in space dimension
$\mathbf{fa}_{sk}$	Feature vector of Fa_k' in space dimension
$\mathbf{W}$	Corresponding coordinate set between $\mathbf{fg}_{si}$ and $\mathbf{fa}_{sk}$

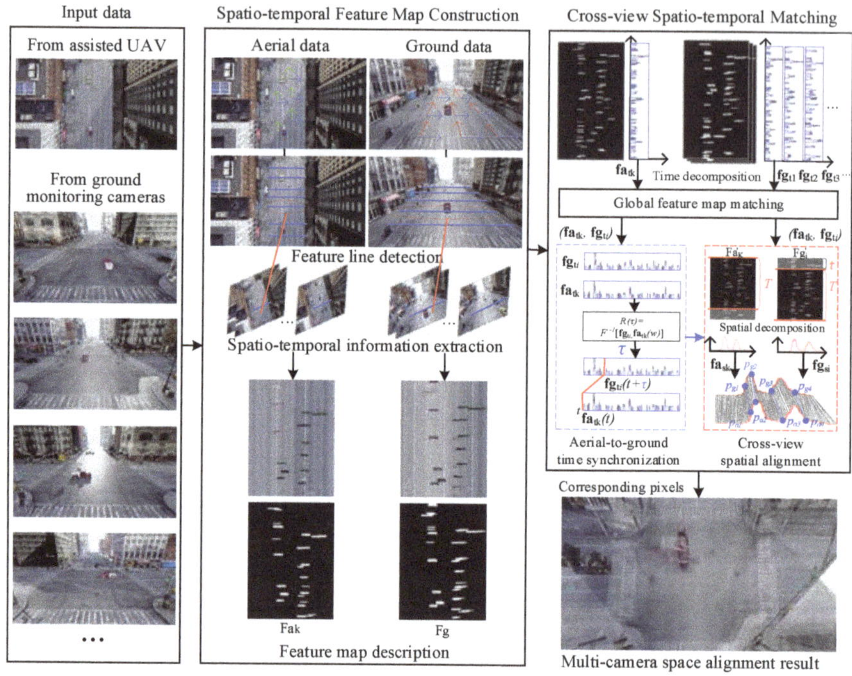

Figure 2. An illustration of the proposed UAV-assisted wide area multi-camera space alignment approach based on spatiotemporal feature map. Our algorithm contains two critical components: spatiotemporal feature map construction to describe the input UAV-assisted aerial data and ground monitoring data and cross-view spatiotemporal matching to mine air-to-ground space correspondences. Multiple ground cameras are aligned into one space with UAV-assisted visual connection.

2.1. Spatiotemporal Feature Map Construction

Before constructing spatiotemporal feature map, we firstly introduce the input data. As shown in Figure 2, the input data contains two parts: ground monitoring videos from ground deployed cameras and aerial videos from the UAV. Among them, each ground monitoring video corresponds to a deployed camera to be aligned. The aerial videos are collected by assisted UAV at different hover positions. The motion information in observation scene is contained in both UAV aided data and ground surveillance data, which is the key to motion consistency for subsequent cross-view matching.

Let N ground monitoring videos $V_{C1}, V_{C2}, \ldots V_{CN}$ denote the monitoring data from N deployed cameras, respectively, and $V_{A1}, V_{A2}, \ldots V_{AM}$ are the UAV-assisted data which are obtained by UAV hovering at M positions. How can these data be described by the spatiotemporal feature map? Similar to the general pipeline of visual feature construction (key point detection, feature extraction and description), our approach consists of three modules: feature line detection, spatiotemporal information extraction and feature map description.

2.1.1. Feature Line Detection

To find spatial correspondences between the UAV data and ground monitoring data, we expect to get the pixel relationship between them for camera space alignment. Therefore, local feature representation method is required for such local information matching. Similar to key point detection in widely-used SIFT algorithm, feature line detection is the beginning step in our proposed spatiotemporal feature map construction method.

What kind of line should we choose as feature line? As is known, there exists a great gap in perspectives between aerial UAV data and ground monitoring data. Perspective projection transformation causes deformation in length, relative proportion and intersection angle. That is why direct scene lines extracted by traditional hand-craft method or deep-learning network are not suitable for air-to-ground matching. However, fortunately, the intersection points of lines are precisely invariant under perspective projection transformation. According to this, we start with the establishment of feature lines. As shown in Figure 2, we draw several lines perpendicular to the direction of vehicle moving as feature lines. This is because such feature lines can capture rich visual intersection information of the moving target passing through them. Thsi intersection information remains unchanged between the air and the ground.

Considering the uncertainty of camera position and orientation, our approach adopts the combination of traffic flow direction and vanish point in [29] to determine feature line. Next, we introduce the proposed feature lines detection method of ground monitoring data and UAV-assisted data, respectively.

For ground monitoring data, feature line directions vary greatly in different camera orientations (Figure 3). When the camera faces the road center (Figure 3a), its feature line directions are quite similar. While for roadside camera, Figure 3b shows an instance of its collected data. The feature line directions are different in different positions. Their included angles are also different in two-dimensional image. Therefore, we need to determine the direction of feature line adaptively according to specific condition. By thoroughly analyzing scene visual information, feature line direction relates to the short edges of foreground moving vehicles. Their slope in different positions is the feature line direction. We adopt vanish point to help extract the feature line direction. This specific method was proposed by Dubská et al. [29], who first obtained the direction of traffic flows by optical flow and calculated the first vanish point by diamond space voting. The feature line corresponds to the direction parallel to the ground and perpendicular to the first direction. Thus, we model background edge to get the edge of foreground moving vehicle. Then, we filter out the edge which belongs to the first vanishing point or perpendicular to the ground. Feature lines are the extensions of these retained foreground edges.

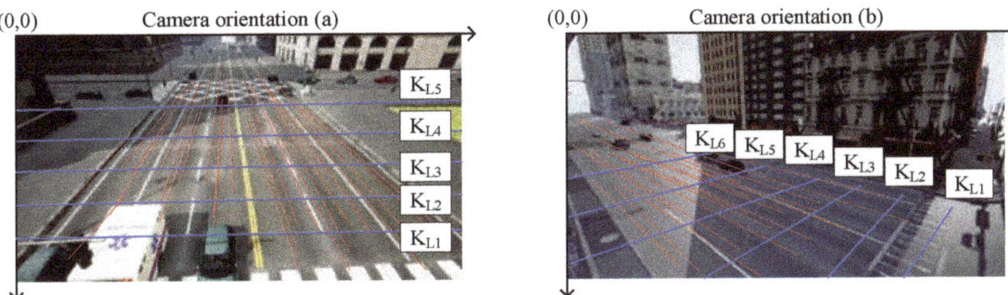

Figure 3. Feature lines in different camera orientations: (**a**) the directions of feature lines are similar to each other; and (**b**) the directions of feature lines are much more different.

As for aerial data from the UAV, its top view makes two-dimensional image without geometric perspective. That is different from ground monitoring data above. Therefore, the feature line direction is just the line perpendicular to the direction of traffic flow in two-dimensional image. Thus, in this part, we only utilize traffic flow detection to obtain feature line. To be specific, the procedure has two steps: traffic flow detection by optical flow approach and feature line drawing with vertical direction of optical flow. It is worth noting that aerial video usually has a wider observation range, which may involve traffic flow in multiple directions. For example, a turning road contains traffic in two directions. Multiple traffic directions correspond to multiple feature line sets. Feature lines in the same direction are grouped into one set.

Based on the methods stated above, we detect and draw feature line in N ground monitoring videos and M UAV-assisted videos. Each video has several feature lines. Taking V as a monitoring video example, it can come from ground cameras or aerial UAV. There are NL_i feature lines detected from V. We finally obtain numerous ground feature lines and aerial feature lines after a series of the above-mentioned processing.

2.1.2. Spatiotemporal Information Extraction

This section aims to extract visual information from input ground monitoring data and aerial assisted data with the help of feature lines. According to motion consistency in cross-view data, we extract the visual information in two dimensions (temporal order and spatial structure). For temporal order extraction, the monitoring video is unframed in order. We record their spatial features from the feature line in turn. Thus, temporal visual feature shown over time can be extracted. Meanwhile, the visual changes in space are the spatial visual feature.

Figure 4 provides the detailed spatiotemporal information extraction method intuitively. Suppose fl_i is one of the feature lines in monitoring video V; it is circled in this figure. V can be ground monitoring camera or aerial camera. The video is decoded into N_i frames at the beginning. The visual data at the position and direction of this feature line can be found at corresponding locations in each frame. Next, the related data are extracted and integrated into one row in order. The number of rows is equal to the number of video frames, which is N_i for fl_i. Figure 4 (right) shows the rows from top to bottom corresponding to the video frames from front to back. The visual data of all rows are the spatiotemporal information extracted from feature line fl_i.

The above visual information extraction approach not only extracts time series information at feature line location but also extracts spatial visual information on the different pixels of feature line. For better understanding, feature lines are similar to a door: the door can obtain what passed by recording what happened in every moment. Similarly, we can get what information go through the feature line by recording visual data in every frame. Thus, the motion time occurred as well as its space position are extracted.

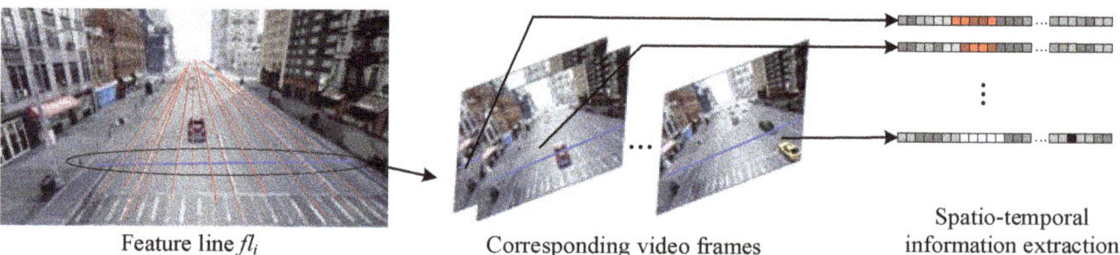

Figure 4. Spatiotemporal information extraction. With the position and direction of a sample feature line (circled on the left), we can extract the visual data at this location from each corresponding video frame. The related visual data are integrated into several rows in order on the right and form the spatiotemporal information.

2.1.3. Feature Map Description

Next, how to describe the above spatiotemporal information is also an important problem. To address this, we construct two dimensional feature map whose coordinate axes are the set of space and time information, respectively.

The spatiotemporal information extracted from the above section is represented by several visual data rows. Based on this, we then connect them in chronological order to form a two-dimensional feature map. The row of visual data calculated in the last section composes one row in the feature map, and the visual data at different time from one feature line position compose one column in the feature map. From the middle module of Figure 2, we can see that the time and position of every passed moving object are recorded in this feature map. The height of each moving object's Y axis in feature map is their passage time through feature line, and the span of each moving object's X axis is object width.

Then, we transform feature map into binary image with small data quantities by foreground object segmentation method. The benefits of this are the following. It can further highlight the motion information which is consistent in different cameras. At the same time, it can also filter out the other visual features that we do not care about (such as color and gradient). The feature map shows that most of the visual data in it are background road. Based on this, we start with the hypothesis that background occupies the majority relative to foreground motion. Then, each column in the feature map is processed as an independent unit to find the background of feature line. According to the statistical distribution of gray value in each column, the gray with maximum value comes from background. Thus, the pixel whose gray value is close to the maximum is set to 255 to indicate background, and vice versa. The binary processed feature map displays obvious black–white effect.

Thus far, the spatiotemporal feature map construction is finished. N ground monitoring videos $V_{C1}, V_{C2}, \ldots V_{CN}$ and M aerial assisted videos $V_{A1}, V_{A2}, \ldots V_{AM}$ are represented by spatiotemporal feature maps. The relationship of videos and their feature maps is one-to-many.

2.2. Cross-View Spatiotemporal Matching

To describe the proposed method clearly, we assume that the aerial spatiotemporal feature map from the UAV is query. To search for its matched database feature map from ground monitoring videos, we propose a cross-view spatiotemporal matching approach which can also determine the best space responding pixel between matched feature map pairs for camera space alignment. The proposed method includes three key steps: (1) global feature map matching; (2) aerial-to-ground time synchronization; and (3) cross-view spatial alignment. The first one measures the similarity of feature maps from the global and the latter two are used to find the corresponding relationship between local pixels.

2.2.1. Global Feature Map Matching

Let Fa_k be the kth query feature map from aerial assist UAV and Fg the database which contains Ng ground spatiotemporal feature maps. It is the collection of ground feature maps calculated from N ground monitoring videos $V_{C1}, V_{C2}, \ldots V_{CN}$, as expressed in Equation (1). Figure 5 gives the whole global feature map matching method.

$$Fg = \{Fg_1, Fg_2, \ldots, Fg_{Nc1}, \overbrace{Fg_{Nc1+1}, Fg_{Nc1+2}, \ldots, Fg_{Nc1+Nc2}}, \ldots, Fg_{Ng}\} \tag{1}$$

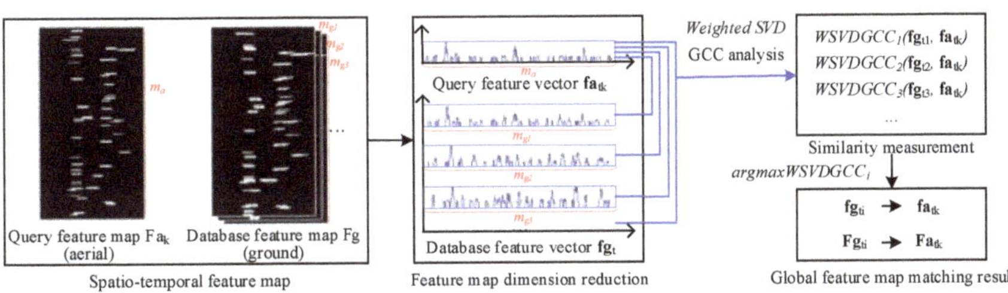

Figure 5. Global feature map matching. Query feature map Fa_k from aerial assisted UAV and database feature map Fg from ground cameras are firstly transformed into one-dimensional time feature vector. Then, we measure the similarity between them according to their weighted SVD generalized cross correlation value (WSVD FS-GCC). The feature map Fg_i corresponding to the highest scoring feature vector $\mathbf{fg}_{ti}$ is the global matching result.

First, the input feature maps Fa_k and Fg are mapped into one-dimensional space before matching. The two-dimensional feature map matching problem is transformed into a one-dimensional feature vector similarity measurement problem. In doing so, it avoids complicated computing accompanied by high-dimensional feature maps while attempts to narrow the gap of 2D feature map caused by air-ground asynchronous. Our method projects 2D feature map to 1D feature vector in time (see Figure 2). To better describe this process, we take feature map $F \in \mathbb{R}^{m \times n}$ as an example. F is m rows and n columns. F can be regarded as several row vectors, as $F = \{\mathbf{r}_1, \mathbf{r}_2, ..., \mathbf{r}_m\}$. The number of row vectors is m and the dimension of each row vector is n. Each row vector comes from a sampling time. Then, as for each row vector, we count the number of foreground pixels as its feature number. Thus, a n-dimensional row vector is converted to a feature number. m row vectors are converted into m feature numbers. By arranging the m feature numbers in order, we can establish the time feature vector of feature map F. Thus, the two-dimensional feature map F, which is m rows and n columns, can be mapped to one-dimensional feature vector $\mathbf{f}_t$, which is an m-dimensional feature vector. The calculation process is shown as follows:

$$\mathbf{f}_t = \{ft_1, ft_2, ..., ft_m\}, \qquad where \quad ft_j = card(\mathbf{r}_p(q) = 0); \quad q = 1, 2, ...n; \quad p = 1, 2, ...m \tag{2}$$

In this way, query Fa_k is represented by $\mathbf{fa}_{tk}$, and all database feature maps in Fg are also represented by time feature vectors. Our next step is to measure the similarity between them. The time non-synchronization problem among air and ground cameras makes the traditional Euclidean distance incapable of quantifying their similarity. This paper analyzes the generalized cross-correlation value between them as the similarity measurement. We adopt the evaluation index of generalized cross-correlation. It was defined by Cobos [30] in 2020, who improved the general generalized cross-correlation based on the sub-band analysis of cross-power spectrum phase, named FS-GCC (Frequency-sliding Generalized Cross Correlation). This method shows robust performance under noise and reverberation. Concretely, according to their denoised FS-GCC values based on weighted SVD, the similarity between query feature vector $\mathbf{fa}_{tk}$ and every database feature vector in $\mathbf{fg}$ can be obtained. In the following calculation, the highest scoring database feature vector is the ground feature vector matched with $\mathbf{fa}_{tk}$. We denote it as $\mathbf{fg}_{ti}$. Meanwhile, their corresponding feature maps Fa_k and Fg_i are a matched pair.

$$i = arg \max_{p}(\mathcal{FS} - \mathcal{GCC}(\mathbf{fa}_{tk}, \mathbf{fg}_{tp})) \qquad where \quad \mathbf{fg}_{tp} \in \mathbf{fg} \tag{3}$$

When all aerial feature maps retrieve their matched ground feature maps in database, we can obtain several feature map pairs, which are the results of global feature map matching. Furthermore, the feature lines corresponding to the same feature map pair are considered as a matched feature line pairs.

After finding the matching relationship between feature maps globally, we next try to find the correspondence between local pixels. The calculation procedure includes two key modules: aerial-ground time synchronization and cross-view spatial alignment.

2.2.2. Aerial-to-Ground Time Synchronization

To find the corresponding pixels between matched feature line pairs, we need to realize time synchronization between them at first. The visual feature is described by spatiotemporal feature maps in this paper, time synchronization and space alignment are closely related. Time synchronization affects the accuracy of finding corresponding points, and then influences the performance of camera spatial alignment. In other words, considering a single variable principle, accurate spatial correspondence is obtained under the prior time unification of spatiotemporal feature maps.

Mathematically, feature vectors $\mathbf{fa}_{tk}$ and $\mathbf{fg}_{ti}$ are one-dimensional time features enriched from the two-dimensional feature maps Fa_k and Fg_i. They are also the time series. The problem of feature maps' time synchronization is also the issue of one-dimensional series' time delay estimation. Generalized cross correlation is one of the most commonly used method. It estimates time delay by analyzing the correlation between two signals. Therefore, our approach employs an improved generalized cross correlation algorithm [30] to synchronize $\mathbf{fa}_{tk}$ and $\mathbf{fg}_{ti}$. This method is used for similarity measurement and global match feature map matching in the previous section. In terms of time delay estimation of $\mathbf{fa}_{tk}$ and $\mathbf{fg}_{ti}$, their concrete time delay τ is the corresponding value when the maximum cross-correlation value obtains. Let $\mathcal{G}$ be the calculation function ($\mathcal{G}$ named $WSVDFC - GCC$ in [30], and we do not bore you with its details), the time delay can be calculated as below:

$$\hat{\tau} = \arg \max_{\tau} \mathcal{G}(\mathbf{fa}_{tk}(t), \mathbf{fg}_{ti}(t + \tau)) \tag{4}$$

$\hat{\tau}$ is the time delay of aerial feature vector $\mathbf{fa}_{tk}$ and ground feature vector $\mathbf{fg}_{ti}$. $\mathbf{fa}_{tk}$ is the reference. The first component of $\mathbf{fa}_{tk}$ and the $\hat{\tau}$th component of $\mathbf{fg}_{ti}$ are synchronized. We reverse $\hat{\tau}$ into the row of feature maps Fa_k and Fg_i, and the collection time of these rows is the same. We than cut the same length T from the synchronization row and get new spatiotemporal feature maps $Fa_k' \in \mathbb{R}^{T \times n_a}$ and $Fg_i' \in \mathbb{R}^{T \times n_g}$. The parameter T needs to meet the following two requirements: $T < m_a$ and $T + \hat{\tau} < m_g$. The specific calculation method is expressed as:

$$Fa_k = \begin{bmatrix} Fa_k' \\ A \end{bmatrix} \quad where \quad Fa_k \in \mathbb{R}^{m_a \times n_a}; \quad Fa_k' \in \mathbb{R}^{T \times n_a}; \quad A \in \mathbb{R}^{(m_a - T) \times n_a} \tag{5}$$

$$Fg_i = \begin{bmatrix} B \\ Fg_k' \\ C \end{bmatrix} \quad where \quad Fg_i \in \mathbb{R}^{m_g \times n_g}; \quad Fg_k' \in \mathbb{R}^{T \times n_g}; \quad B \in \mathbb{R}^{\hat{\tau} \times n_a} \in \mathbb{R}^{(m_g - T - \hat{\tau}) \times n_a} \tag{6}$$

2.2.3. Cross-View Spatial Alignment

With Fa_k' and Fg_i', we next solve the problem of finding corresponding pixels for cross-view spatial alignment. Our proposed method includes three steps: (1) feature map dimension reduction; (2) one-dimensional space feature vector alignment; and (3) cross-view air-to-ground spatial alignment.

Similar to Section 2.2.1 that maps feature map to time dimension, we map $Fa_k' = \{\mathbf{ca}_1^T, \mathbf{ca}_2^T, ..., \mathbf{ca}_{n_a}^T\}$ and $Fg_i' = \{\mathbf{cg}_1^T, \mathbf{cg}_2^T, ..., \mathbf{cg}_{n_g}^T\}$ to space dimension at first. Figure 6 displays the proposed feature dimension reduction and alignment process vividly. As we can see, Fa_k' and Fg_i' are reduced to one dimension as space feature vectors $\mathbf{fa}_{sk}$ and $\mathbf{fg}_{si}$. The length of space feature vector is equal to the number of columns in feature map and each component is the number of foreground pixels in the corresponding column. The calculation formula is as follows.

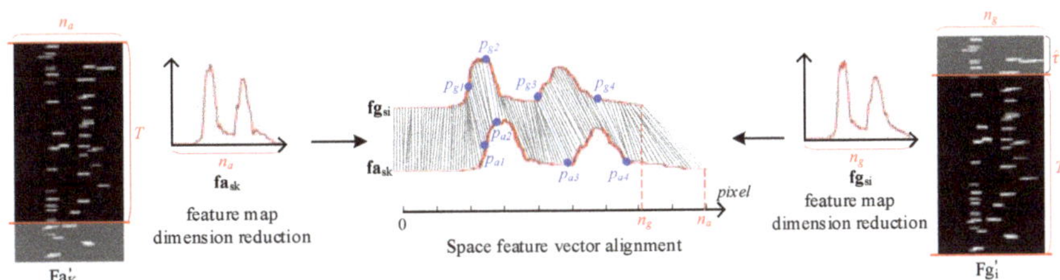

Figure 6. Cross-view spatial alignment. The matched feature map Fa_k' (**left**) and Fg_i' (**right**) is firstly reduced from 2D map to 1D spatial feature vector, as denoted by $\mathbf{fa}_{sk}$ and $\mathbf{fg}_{si}$. We then match the two spatial sequences by DTW (**middle**). Several corresponding pixel pairs labeled in blue are returned as result.

$$\mathbf{fa}_{sk} = \{fa_{sk1}, fa_{sk2}, ..., fa_{skn_a}\} \quad where \quad fa_{sk_p} = card(\mathbf{ca}_p(q) = 0); \quad q = 1, 2, ...T; \quad p = 1, 2, ...n_a \tag{7}$$

$$\mathbf{fg}_{si} = \{fg_{si1}, fg_{si2}, ..., fg_{sin_g}\} \quad where \quad fg_{si_p} = card(\mathbf{cg}_p(q) = 0); \quad q = 1, 2, ...T; \quad q = 1, 2, ...n_g \tag{8}$$

Note that the length of space feature vectors are different because the different sizes of feature maps.

We leverage Dynamic Time Warping (DTW) [31] as the matching method to align a pair of space feature vector $\mathbf{fa}_{sk}$ and $\mathbf{fg}_{si}$. It is a simple but effective template matching algorithm which is also universality for different sequence lengths. Our first stage is to construct a distance matrix $D \in \mathbb{R}^{n_a \times n_g}$. $D(x,y)$ is the Euclidean distance between the xth element of $\mathbf{fa}_{sk}$ and the yth element of $\mathbf{fg}_{si}$. After that, we start to align the two sequences. The matching path is set as $\mathbf{W} = w_1, w_2, ..., w_j, ..., w_l$ $(max(|n_a|, |n_g|) \leq l \leq |n_a| + |n_g|)$. Each element $w_j = (x, y)$ represents the aligned coordinate pair (xth coordinate of $\mathbf{fa}_{sk}$ aligns with the yth coordinate of fg_{si}). To ensure each element in the sequence can find its corresponding alignment position without intersection, W needs to satisfy:

$$w_1 = (1, 1) \tag{9}$$

$$w_l = (n_a, n_g) \tag{10}$$

$$w_{j+1} = (x', y') \qquad x \leq x' \leq x+1 \quad y \leq y' \leq y+1 \tag{11}$$

where x' and y' are the next matched coordinates of $\mathbf{fa}_{sk}$ and $\mathbf{fg}_{si}$. It only has three possible results: $(x+1, y)$, $(x, y+1)$, $(x+1, y+1)$. We choose the one with the minimum cumulative distance from (x, y) according to distance measurement D. After that, we can obtain the matching relationship between vector elements which is stored in $\mathbf{W}$. However, there are diversified corresponding relation types which include one-to-many relationship, many-to-one relationship and one-to-one relationship. The first two are ambiguous in spatial alignment, so we only retain the one-to-one matching pixel pairs. At the same time, we further sample these one-to-one pixel pairs at equal space intervals to get sparse space correspondences. After such screening, $\mathbf{W}'$ is the corresponding coordinate set between $\mathbf{fa}_{sk}$ and $\mathbf{fg}_{si}$.

Feature maps Fa_k and Fg_i constructed by $\mathbf{fa}_{sk}$ and $\mathbf{fg}_{si}$ are just an example of matched feature map pairs. All feature map pairs calculated after Section 2.2.1 can obtain their corresponding relationship between local pixel by the methods in Sections 2.2.2 and 2.2.3. Thus, several corresponding coordinate sets are returned. Moreover, we can track back to the feature line and camera corresponding to each set. This means that we obtain several cross-view corresponding points between aerial 2D images captured by assisted UAV and ground 2D visual data collected by deployed monitoring cameras.

Once air-to-ground corresponding pixels are matched, we calculate the homography matrix between cameras by more than four non-collinear corresponding coordinate pairs. The relative projection relationship between them can be estimated. In this way, the proposed method gets the relationship between each ground monitoring camera and the assisted UAV. The M locations of assisted UAV can be united into one coordinate system with current visual positioning and navigation methods (e.g., SLAM), so ground deployed cameras are aligned to this coordinate system naturally. Our system realizes multi-camera space alignment in large scale environment under UAV assistance.

3. Experiments

We conducted extensive experiments to evaluate the performance of our proposed multi-camera space alignment approach based on spatiotemporal feature map. To maintain the objectivity and comprehensiveness, we constructed an evaluation database by ourselves, which is described in Section 3.1. On this basis, we then explored the robustness and accuracy of our proposed method from both qualitative and quantitative aspects in simulation environment and real scene. The extended applications of our approach are provided in Section 3.4.

3.1. Database

Database in simulation environment

This paper utilizes AirSim [32] as the simulation platform to construct a suitable virtual scene for our system's performance verification. AirSim is an open source simulator based on Unreal Engine. It supports cross-platform operation, multiple programming languages and various sensors (camera, UAV, Lidar, GPS, etc.). Some major parameter settings in Airsim are summarized in Table 2, including environmental parameters and sensor parameters. Figure 7 presents the simulation scene model and some simulation monitoring data.

Table 2. The parameter settings to generate database in simulation environment and real scene.

	Environmental Parameter		**Sensor Parameter**	
Simulation Environment	Environment intensity	1.0	Ground camera number	11
	Directional light actor	light source	Ground camera resolution	1920 × 1080
	Colors determined by sun position	Yes	Ground camera FOV	90°
	Sun brightness	75	Aerial camera position	5
	Sun height	0.348239	Aerial camera resolution	1920 × 1080
	Horizon Falloff	3.0	Aerial camera FOV	90°
	Diffuse boost	1.0	Acquisition frame rate	25 fps
Real Scene	Scene type	Mixed traffic system	Ground camera number	4
	Acquisition time	15:00 p.m.	Ground camera resolution	1920 × 1080
	Scene width	≈60 m	Aerial camera position	1
	Scene length	≈50 m	Aerial camera resolution	1920 × 1080
	Ground camera height	≈7 m	Aerial camera FOV	58°
	UAV flight altitude	≈80 m	Acquisition frame rate	25 fps

To be specific, we chose a model of urban street block as our simulation environment, as shown in Figure 7a. It includes abundant and complex city elements: buildings, landscape plants, traffic signs, junctions, etc. Based on this model, we firstly load multiple car models and set various running routes to restore the real traffic flow as much as possible. Then, the camera model and UAV at different positions are added to imitate ground monitoring cameras and aerial auxiliary camera. Thereafter, we collect simulation monitoring data with these cameras and establish a test simulation dataset called $CamData - Sim$. This database consists of two parts: (1) 24 videos from 11 ground cameras at fixed locations; and (2) 5 aerial videos from the UAV at 5 hover positions. Their frame resolution and rate are set to 1920 × 1080 and 25 fps, respectively. The self-built simulation database $CamData - Sim$ is provided in Figure 7 (right). Several vehicles shuttle through these streets and their moving information is collected into ground monitoring videos and the UAV videos independently. Moreover, to better evaluate the effectiveness of the proposed method, these videos are captured by ground cameras and UAV at different heights with different pitch angles.

Database in real scene

Taking into account that current public multi-camera databases cannot provide both ground monitoring data and auxiliary UAV data, we constructed a new multi-camera monitoring database. Figure 8 provides the collection environment and data of our self-built database. Table 2 provides its related parameter settings, in which some cannot be obtained in real scenes and only roughly estimated parameters are given. (1) Acquisition environment: This database is collected from a mixed traffic system with bidirectional six-lane main road and bidirectional four-lane side road. The width of its middle green belt is about 25 m and the total transverse length of this road is more than 60 m. (2) Camera configuration: There are four ground monitoring cameras and each camera monitors traffic in one traffic area, including northbound main road, northbound side road, southbound main road and southbound side road. Since the accurate parameters of a deployed multi-camera system are unknown, we make a rough estimation of its main parameters. The deployment heights of ground cameras are about 7 m and their pitch angle is about 60°. As shown in Figure 8, there is little overlap between their field of view. The auxiliary UAV data which overlook the observation scenario are captured at about 80 m. It contains common motion information with the ground monitoring data. (3) Data information: The frame resolution and rate of all data in this database are 1920×1080 and 25 fps, respectively. The total frame number of each ground video is 5493. The time delay and space relationship between these data is unknown. This paper applies the proposed approach to estimate the space alignment relationship between the ground cameras with UAV assistance.

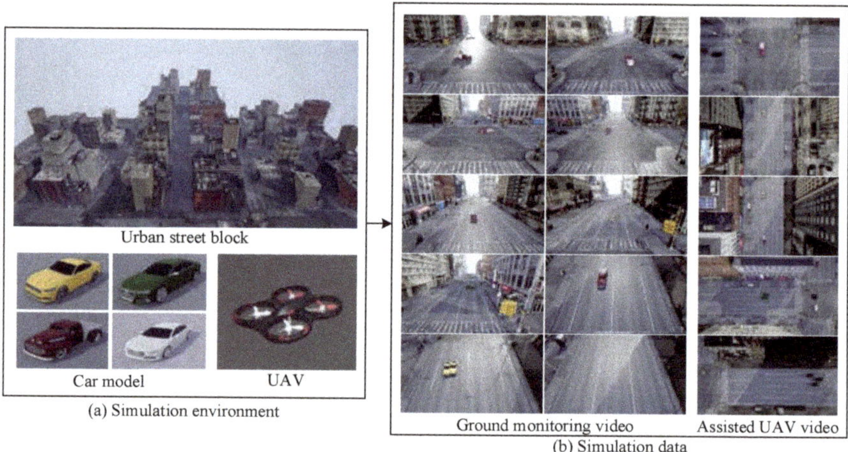

Figure 7. Our simulation database . (**a**) Simulation environment. The top left figure is a model of urban street block. The car models and UAV used in database are displayed below. (**b**) Some examples of ground monitoring videos and assisted UAV videos in this simulation database.

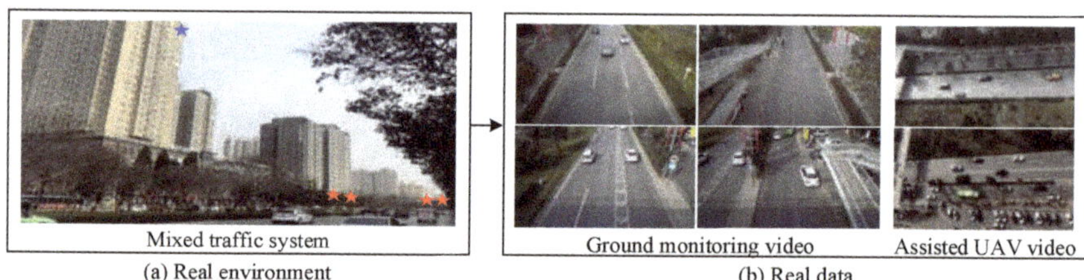

Figure 8. Our self-built database in real scene. (**a**) Real environment. It is a mixed traffic system with bidirectional six-lane main road and bidirectional four-lane side road. The stars represent UAV and ground camera's general locations. (**b**) Real data. the ground monitoring videos and the UAV videos captured from real scene.

3.2. System Performance Evaluation on Simulation Environment

In this section, we explore the performance of our proposed approach on three typical traffic scenarios: crossroad, T-junction and straight road. In addition to qualitative analysis, we also conduct quantified analysis on simulation environment in which the ground truth is manually labeled. Figure 9 displays some space alignment results and Table 3 the pixel error statistics.

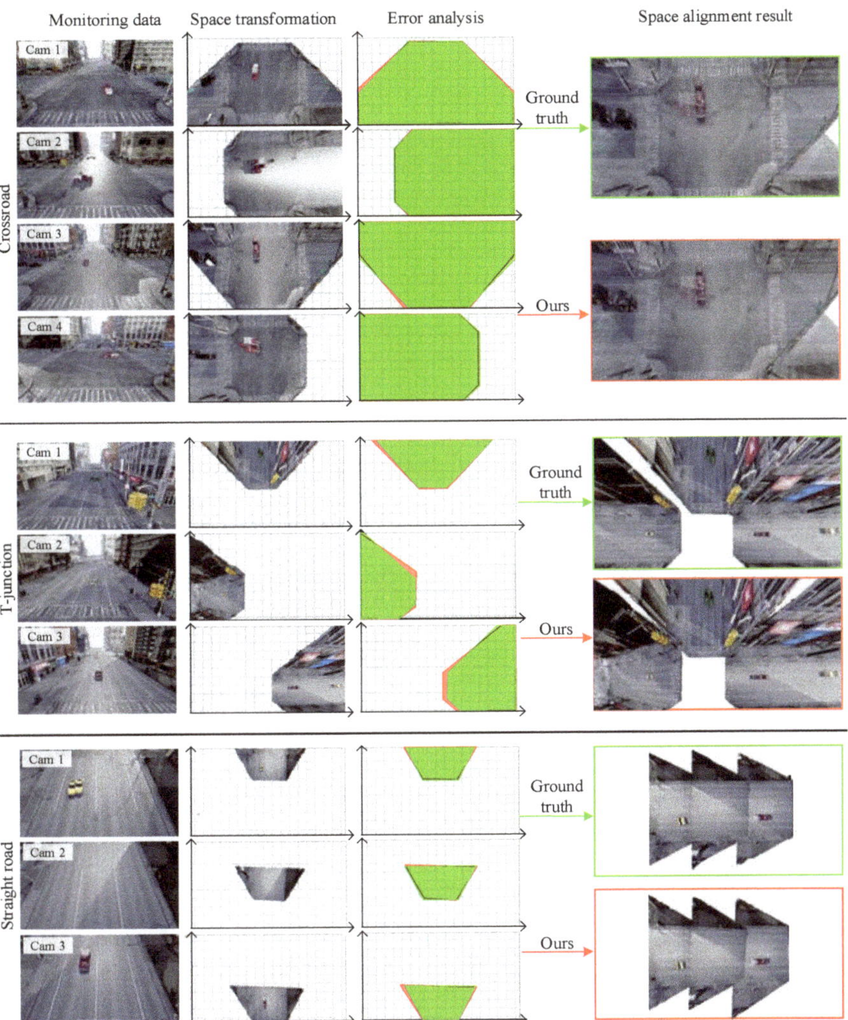

Figure 9. Some qualitative evaluation results of our proposed approach on simulation environment. Three groups of experiments conducted on crossroad, T-junction and straight road are displayed from top to bottom, respectively. Their space alignment results are shown in turn, including the monitoring data, space transformation and the comparison with ground truth. In these results, the ground truth is marked in green and our results are marked in red.

Figure 9 shows the space alignment results of crossroad, T-junction and straight road, from top to bottom, respectively. The fist column is the monitoring data of ground deployed cameras. They are transformed into the united coordinate system with the alignment parameter. The following error analysis roughly evaluates algorithm performance by the coincidence degree between ground truth (marked in green) and our results (marked in red). Multi-camera space alignment results are shown in the end. It can be seen in this figure that our approach performs well in different situations. The first simulation scene is a crossroad. Four overlapping monitoring cameras monitor traffic from four directions. To better see the space alignment performance of overlapping region, the space alignment results of crossroad are set to translucent. Compared with ground truth, we can find that zebra crossings are mapped together successfully. That means the same visual information is aligned to the same coordinates, which indicates the effectiveness of our approach. The second simulation scene is a T-junction and the ground monitoring cameras in it have limited overlapping area between them. Under UAV assistance, these three cameras are calibrated into one space. This illustrates that our system can maintain stable performance with partial-overlapping cameras. The bottom test situation is a straight road with three cameras of sequential distribution. Their overlapping region is not only limited but has fewer visual features. As the right column shows, we can return good space alignment results, further verifying the robustness of our proposed method.

The quantitative experimental study was conducted by analyzing the pixel error between our space alignment results and ground truth. Table 3 shows that the pixel error varies from 5.78 to 23.76 pixels. The average errors on above three scenarios are 20.02, 20.32 and 10.01 pixels, respectively. Thus, if we want to relate the visual data of different cameras, the space alignment error is within 25 pixels. This set of evaluations on different monitoring scenes further demonstrates that the proposed approach satisfies the need in the practice interconnection application. Meanwhile, these quantitative results are also in good agreement with the previous qualitative results. In addition, we can see in this table that there are some differences of space alignment error between different scenarios. The performance of straight road is better than that of crossroad and T-junction. The reason for this phenomenon is as follows. Crossroad and T-junction have both turning and straight traffic. They include more complex motion compared with straight road. This leads to more disturbances of feature line detection and spatiotemporal feature map construction, which directly influences space alignment performance.

Table 3. The pixel error of different monitoring scenarios.

Scene	Crossroad					T-Junction				Straight Road			
Camera	1	2	3	4	AVG	1	2	3	AVG	1	2	3	AVG
Pixel error	23.76	16.33	23.71	16.29	20.02	22.09	17.23	21.65	20.32	14.11	10.13	5.78	10.01

Overall, the evaluation in a simulation environment shows that our proposed multi-camera space alignment approach obtains satisfactory performance not only in quality but also in quantity.

3.3. System Performance Evaluation on Real Environment

Besides evaluation on simulation environment, we also evaluated the performance of our system in a real environment. The test scene and monitoring data constructed by ourselves is introduced in detail in Section 3.1. We applied the proposed method to align the four ground monitoring cameras into one united coordinate system.

Figure 10 shows our space alignment result in real traffic scene. The monitoring data from four ground cameras are mapped into a united coordinate system, as shown in the second column. We then compare our result (labeled in red) with ground truth (labeled in green by manual calibration) qualitatively. The comparison results of the individual camera and the whole system are both provided. Viewing the result as a whole, we can see that these ground cameras are well aligned. Their space alignment results replay the whole monitoring scene, which is a bidirectional traffic system with greenbelt. The imaging relationship between these limited overlapping ground cameras can be obtained with UAV connection. This means that cameras can cooperate for overall surveillance. By comparing with the ground truth, the pixel error of our approach is about 20, which demonstrates the feasibility and effectiveness in real environment. From a local point, lane direction after each camera mapping is basically parallel. That conforms to the actual situation, which also confirms algorithm performance. However, as we can see, our approach performs poorly on the distant targets which are warped incorrectly with too large longitudinal extension.

This happens because our method cannot get enough feature lines when the object is too small in the far region.

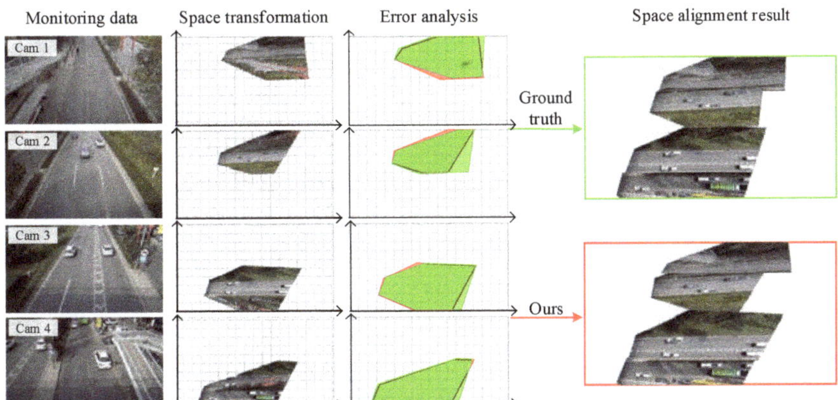

Figure 10. Some qualitative evaluation results of our proposed approach on real environment. The real monitoring scene is two-way multi-lane traffic system with main and side road. Our space alignment results are shown in turn, including the monitoring data, space transformation and the comparison with ground truth. In these results, the ground truth is marked in green and our results are marked in red.

The above experiments were conducted on a computer with an Interl(R) Core(TM) i9-9900X (3.50 Hz GPU, NVIDIA GeForce GTX 1080Ti GPU, 64 GB RAM) using C++. The computational complexity of the proposed method is analyzed below. As described, the proposed method contains two main modules: spatiotemporal feature map construction and cross-view spatiotemporal matching. For the real scene above, the running time of the first module is about 123.9 s. In the second module, the feature map dimensional reduction in time and space costs 38.2 s on average. The time of air-to-ground time synchronization and cross-view spatial matching is 136.0 s. To sum up, the time cost to align the ground monitoring cameras in the above real scene is within 5 min. That also verifies the fast spatial alignment ability of our system in large scenes. In addition, the proposed method is easier to operate in real environment. The operation complexity mainly comes from input data preparation. The input ground monitoring videos can be obtained from database or real-time monitoring data. The UAV needs to capture the monitoring space from top view under stable flight condition. We only need a part of the common motion information between ground and aerial data and do not require data synchronization.

3.4. Extended Applications

Due to its multi-camera space alignment ability, the proposed method has great value in many real-world scenarios. For example, vehicle road hybrid system is a common traffic scene. Multiple cameras are used in it to monitor traffic operation status. The proposed method can be applied to estimate the space relationship between cameras and converts independent monitoring to integrated monitoring. The efficiency of traffic monitoring can be improved. A campus is a typical example of our approach's application scenario. To insure teachers and students work or study on a harmonious campus, many cameras are deployed in every corner of the campus. The proposed method can be used to obtain the spatial position of each camera in a campus and unify them into a coordinate system. Thus, all monitoring data will be aligned as a whole. We can see what is happening on campus from the whole multi-camera video rather than multiple separate single camera videos. Besides the above two examples, our approach also can be applied to key industrial factories, large-scale activity square, etc.

In addition, the proposed method, which lays the foundation of multi-camera system, has the potential application value in many multi-camera cooperation fields, including object re-identification, multi-object detection, multi-camera cooperative locating, and so on. To be specific, on the basis of our multi-camera space alignment results, cross-camera object re-identification can be solved from a new perspective. Other object re-identification methods identify the same target by their feature similarity. Unlike other methods that mine their similarity, we can relate the same object across different cameras by the estimated spatial corresponding relationship. Furthermore, the initial object detection result can be verified by multiple cameras with their space alignment result. Through the spatial correspondence between target boxes, false alarm rate and missed rate can also be reduced. For multi-camera cooperative locating, their space alignment result can provide a references location of the interested target. Especially, when the target is occluded, the result obtained by our approach can ensure stable positioning accuracy.

To show the utility of our proposed approach in real-world intuitively, we apply it in a typical vehicle road hybrid system. Figure 11 shows a crossroad with four ground monitoring cameras. They observe the traffic intersection from four directions. The proposed method has application value in imaging display and intelligent analysis. Concretely, on the one hand, the proposed method aligns the four cameras into one coordinate system. Four independent monitoring videos are unified into a more comprehensive monitoring video, as shown on the right. That allows users to timely obtain the whole intersection running state, which improves the efficiency of current video surveillance. On the other hand, the spatial correspondence between different cameras obtained by our approach also contributes to cross-camera intelligent analysis. If we employ single camera object detection algorithm on one of these cameras, the objects can be detected out. As shown in the upper left corner, a white car marked with red box is detected out. On the basis of our result, the data of this target in other cameras can be directly associated. In other words, such ability to relate targets across cameras is capable of cross-camera re-identification and tracking. Compared with other methods which detect objects in different cameras separately first and then re-identify them, our approach only detects objects of one camera and relate them by coordinate correspondence. The efficiency and robustness of cross-camera intelligent analysis are naturally improved.

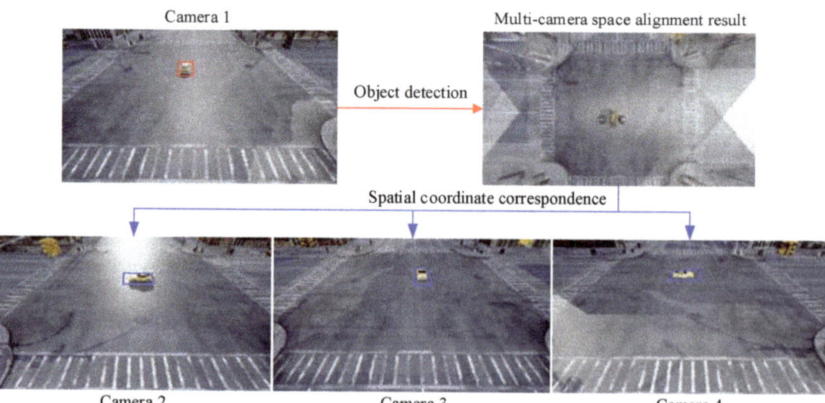

Figure 11. An applicable example of the proposed multi-camera space alignment approach. Our method aligns the four ground monitoring locations into one coordinate system. With this corresponding spatial relationship, the object information of the car in Camera 1 (marked in red) can be directly associated with its data in other related cameras (marked in blue).

4. Discussion

4.1. Performance Comparison

In this section, we compare our performance with other works from two levels. First, from the partial important sub-process, we conducted contrast experiments on cross-view matching performance, which is one of the key technologies in our system. Then, from the overall performance, we compared our approach with other methods on multi-camera space alignment.

4.1.1. Comparison of Cross-View Matching

Cross-view matching, which mines the relationship between ground monitoring camera and auxiliary UAV, is one of the key technologies involved in this paper. Its accuracy has a direct impact on air-to-ground coordinate system unification, and thus plays a negative role in multi-camera space alignment. Therefore, first, the performance our proposed cross-view matching algorithm is compared with other matching methods in both simulation and real environments.

SIFT (Scale-Invariant Feature Transform) [33] proposed by Lowe and SuperGlue [34] proposed by Sarlin are chosen as the contrast methods. SIFT as a traditional hand-crafted matching approach that is widely used in practical application. It extracts local feature from input image and measures them similarity by Euclidean distance. The highest scoring feature and query feature are the matching pair. SIFT is robust to rotation, zoom scale and brightness changes. SuperGlue [34], as a deep neural matching network, was recently proposed. It is based on graph neural network and attention mechanism. They regard matching as the optimal transport problem in which the loss function is constructed by deep network. In the specific implementation, two images and their visual features described by SuperPoint [35] are the input. They are then sent to the matching network established by SuperGlue and the matching relationship between them is returned as output.

Figure 12 shows the qualitative performance comparison of SIFT, SuperGlue and our approach on cross-view matching. The test image on the left is a UAV aerial image, and its related ground monitoring image is provided on the right. They observe the monitoring scene from the top view and street view, respectively. Obviously, there exists great perspective gap between them. The evaluation results on simulation environment and real traffic scene are provided from top to bottom.The above three methods are applied on the two scenes for cross-view matching. For visualization, the matching pairs found by each method are connected with straight lines. We can see that our proposed method outperforms the other methods in both quantity and accuracy. (1) For quantity, our approach returns more than 60 matching pixel pairs. SIFT only obtains a few matching pairs. SuperGlue finds plenty of matching pairs in the simulation environment, but it finds very few pairs in real scene. (2) For accuracy, most of the matching pairs calculated by SIFT are not correct. Similarly, SuperGlue can also hardly find the accurate cross-view corresponding point. However, in the matching results of our system in the simulation and real environments, the overwhelming majority of pairs are accurate. According to the above analysis, SIFT gets too few and incorrect matching pairs. The accuracy of SuperGlue is also poor on air-to-ground matching. In other words, the two approaches fail on cross-view matching. However, our proposed method can obtain sufficient and correct matching pairs. It shows satisfying performance across different perspective views.

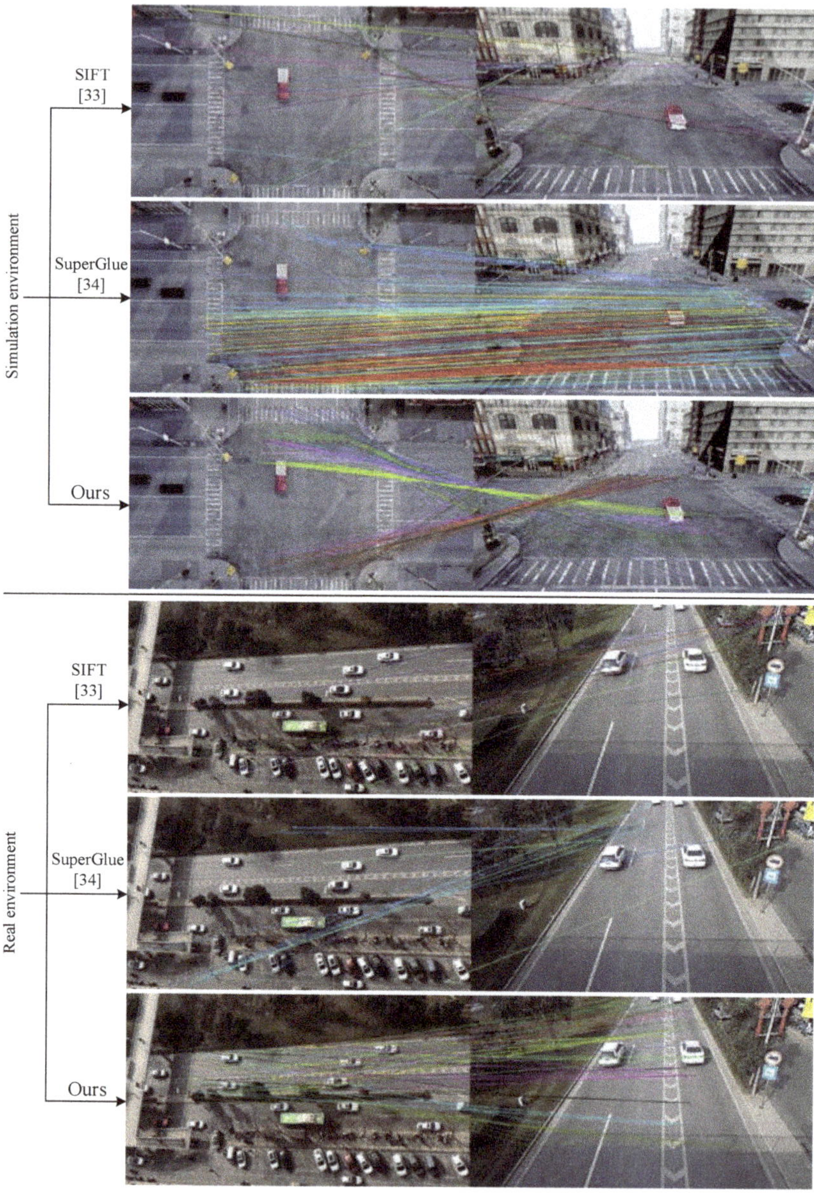

Figure 12. Qualitative cross-view matching comparison of our proposed method against SIFT and SuperGlue. The air-to-ground matching pairs between UAV aerial image and the ground monitoring image are connected by straight lines.

The high performance of our approach is due to the proposed spatiotemporal feature map and cross-view matching method, which links up different views according to intersection invariance of projection transformation. Thus, it is naturally robust to view change. However, SIFT matches images by their local feature similarity. That makes it difficult to cover such huge view gap. Meanwhile, there are many similar elements in the observation scene, e.g., the pedestrian crossings in four directions. That is also a key reason for the poor performance of SIFT. SuperGlue is based on a pre-trained matching network. Its performance depends on the scale and quality of the training database. The disadvantage of matching neural network on generalization capability causes its failure of cross-view matching. To sum up, experimental evaluation and result analysis prove that our approach has better performance than the comparison methods in cross-view matching.

4.1.2. Comparison of Multi-Camera Space Alignment

For the overall multi-camera space alignment performance, we compare the proposed method with other two methods: COLMAP and MapNet.

COLMAP is a widely used 3D reconstruction approach based on structure-from-motion [36] and multi-view stereo [37]. Without camera calibration in advance, COLMAP can reconstruct the whole scene with a set of ordered or unordered two-dimensional images. For multi-camera space alignment, we use COLMAP to reconstruct the whole monitoring scene by inputting a series of scene images obtained from different angles. Then, the monitoring data from multiple ground camera as the new registered images can be re-localized into scene reconstructed model. Thus, multiple cameras are united into the coordinate system established by scene reconstructed model. Thus, COLMAP can also achieve multi-camera space alignment based on three-dimensional reconstruction.

MapNet [38] is a camera localization approach with geometry-aware learning of maps. It was proposed by Brahmbhatt in 2018. In this work, they proposed a novel parameterization method for camera rotation to better estimate camera pose with deep learning network. In other words, MapNet can be regarded as an end-to-end multi-camera method. The ground monitoring data can be sent to this network and the output is each camera's pose in the whole scene map. The relative space relationship is also contained in their poses. On the basis of multiple camera localization, we can align them. Therefore, MapNet can realize multi-camera space alignment by camera localization.

As described above, COLMAP and MapNet are not proposed for multi-camera space alignment. The reason that we choose them as the comparison methods are as follows. First, the related works for overlapping cameras and non-overlapping cameras are not suitable for comparison. The overlapping relationship between cameras in wide area multi-camera system is usually chaotic and unknown. Marker- or motion-based methods can estimate camera spatial topological relations and not the pixel level correspondence. Second, COLMAP as a representative algorithm of 3D reconstruction and MapNet as a deep learning method can obtain camera space relationship in some ways. They can implement multi-camera space alignment with data post processing. The comparison with them can reflect the performance of our method on space alignment.

The multi-camera space alignment results of the above three methods are displayed in Figure 13. As we can see, the test scene is a crossroad with four ground monitoring cameras. From left to right, the results of COLMAP, MapNet and ours are provided. COLMAP successfully aligns the monitoring data captured from four cameras into one coordinate. The same visual information (e.g., the zebra crossings) is mapped with the same two-dimensional coordinate. As for MapNet, it fails to align all monitoring data into one united coordinate system. Especially, the result of Camera 2 maps the data into the wrong coordinates. That leads to a large pixel error with manually labeled ground truth. Meanwhile, the final alignment result is also formless, which makes it hard to monitor the scene in all directions. The alignment result obtained by our proposed method shows comparable qualitative performance with COLMAP. The four monitoring cameras are also well aligned into one united coordinate system. We can see that the error between ground truth and our result is very small. To quantitative compare the pixel error, we statistically analyze the error of each method, as shown in Table 4. It is the qualitative experiment results. COLMAP obtains the minimum pixel error on each camera space alignment, while MapNet has quite large pixel error. The pixel error of our approach is about 20 pixels, which can meet the demand in real-world.

Table 4. The quantitative comparison of COLMAP, MapNet and ours on pixel error.

	Camera 1	Camera 2	Camera 3	Camera 4	AVG
COLMAP [36,37]	8.25	3.84	9.89	5.88	6.965
MapNet [38]	174.61	88.34	59.59	231.49	138.51
Ours	23.76	16.33	23.71	16.29	20.02

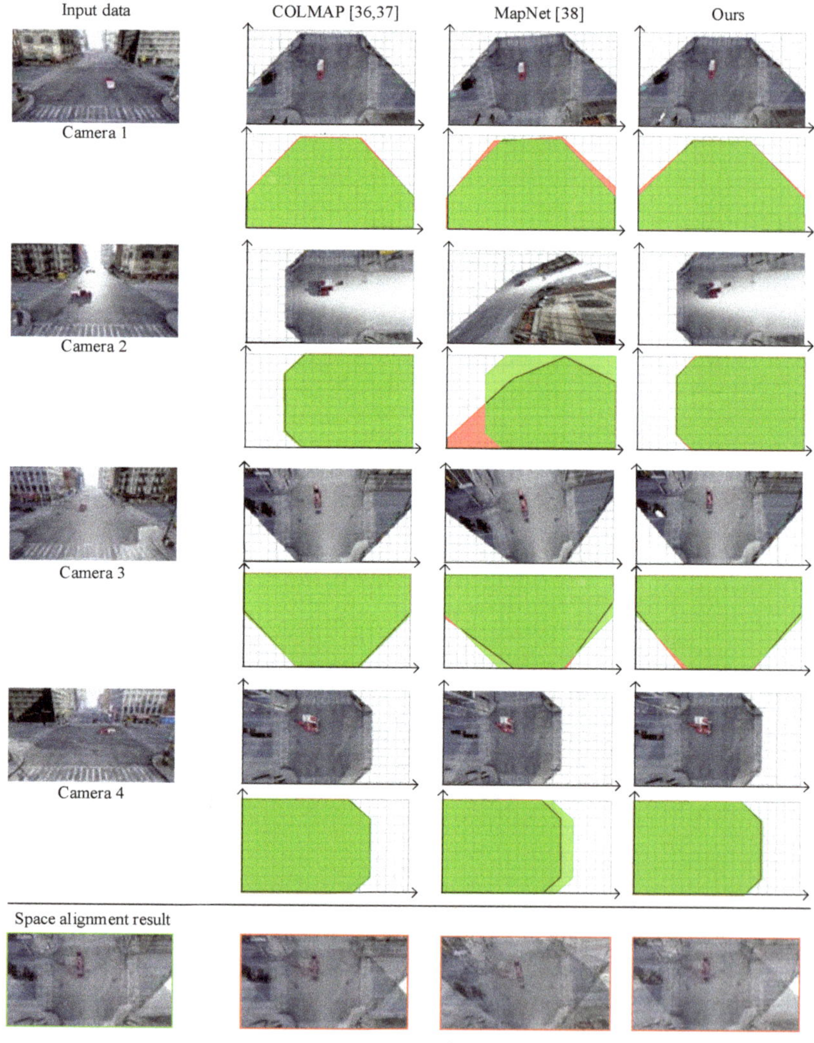

Figure 13. The comparison of multi-camera space alignment performance. The test scene is a crossroad with four monitoring cameras. The coordinate mapping results of each camera by COLMAP, MapNet and ours are provided from left to right. The space alignment result in the bottom marked in green is the ground truth.

The factors causing the above results are analyzed below. The high performance of COLMAP is due to the space relationship provided by its pre-built scene 3D model. A better scene model guarantees accurate space alignment. However, such scene model is usually obtained by a variety of scene images, requiring 10 h for three-dimensional reconstruction. With the increase of the number of cameras and monitoring area, it will take more time. It cannot meet the needs of fast spatial alignment in large scenes. The performance of MapNet is limited by the deep neural network. It can regress camera pose by multi-layer network computing and pre-data training. However, such pose regression method still has accuracy disparity with the method based on geometry structure and image retrieval. For the proposed method, it ensures space alignment efficiency with the help of UAV, which has excellent flexibility and global awareness that can adapt to the needs of fast spatial alignment in large scenes. Meanwhile, we mine the motion consistency between UAV and ground monitoring cameras. Thus, we can align them into one united coordinate system by air-to-ground pixel correspondence. That ensures the space alignment accuracy. To balance the efficiency and accuracy, our approach returns better performance than the other contrast methods.

4.2. Parameter Discussion

This section discusses the effect of three parameters on our system's performance: the number of feature lines, camera pitch angle and deployment height. The number of feature lines relates to cross-view visual feature extraction and description. Different camera pitch angles and deployment heights are also two main factors influencing our performance.

The evaluation data are captured from a typical crossroad on simulation environment, as presented in the top of Figure 9. To simulate different situations, we vary the view angle and deployment height of ground cameras. Meanwhile, the number of feature lines is also changed to analyze algorithm performance. Using variable-controlling principle, the pixel errors by varying these parameters are studied. The experimental results are given in Figure 14. It provides the proposed method's pixel error under different camera pitch angles ($20°$, $30°$ and $40°$) and different camera deployment heights (5 and 9 m) with different number of feature lines (the range interval is $[50, 600]$).

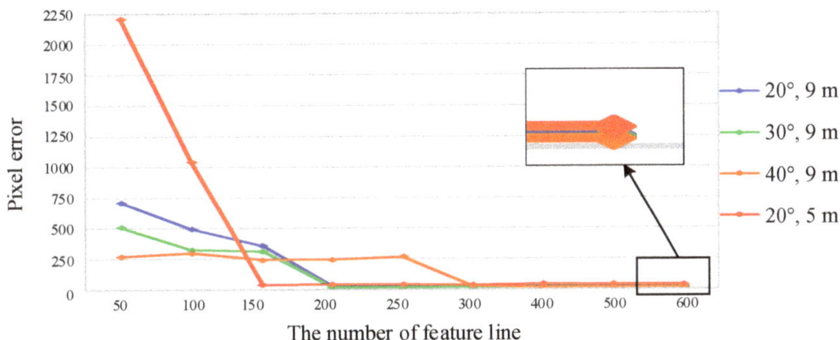

Figure 14. Pixel error under different camera pitch angle, deployment height and the number of feature lines. Blue, red and orange curves are the performance of different pitch angles with different number of feature lines based on the same deployment height of 9 m. Blue and green curves are the performance of different deployment with different number of feature lines based on the same pitch angle 20°.

First, the error decreases with increasing the parameter of number of feature lines, as shown by the three curves. When this number is large enough, the algorithm error keeps at a low level. That is because more feature lines mean more spatiotemporal feature maps. The input data can be described more comprehensively, and then rich cross-camera corresponding points can be obtained. Thus, the accuracy is greatly improved at the beginning. However, when we have enough features and corresponding points, the error will not be greatly reduced. An appropriate number of feature lines for the proposed method is about 200–300. At the same time, it can be seen that the greater is the camera pitch angle, the better is the performance. Overall, $40°$ obtains the minimum alignment error. Large pitch angles of the ground monitoring camera have a small perspective gap between it and the aerial UAV. That makes air-to-ground matching more accurate and further improves multi-camera space alignment performance. In addition, we also found that the situation with $40°$ pitch angle converges to the minimum error more slowly than others. Under the same deployment height, the more the camera looks down, the smaller its observation range is. Therefore, it requires more feature lines for accurate space alignment.

The blue and green curves show the impact of different deployment heights on alignment error. We enlarge the error results after convergence in the upper right corner. It is noticeable that the space alignment error of cameras deployed at 9 m is lower than those at 5 m. It is for the same reason that a large pitch angle has a smaller error. High deployed cameras have more similar perspective views with auxiliary UAV. They can be aligned into the united coordinate system established by UAV more accurately. The change regularity of feature line number also verifies the discussion in the previous paragraph.

However, there are two major limitations to this study that will be addressed in the future. First, the proposed multi-camera space alignment approach is based on UAV-assisted aerial data, which unifies ground monitoring cameras. Thus, it is not applicable to these monitoring situations where stable UAV video cannot be obtained, e.g., no fly zone for UAV, bad weather so the UAV is unable to hover stably or areas that are covered by trees or other things. Secondly, the performance of our proposed method depends on the spatiotemporal feature map which describes input data with abundant traffic flow. However, it is affected by random traffic flow. When the passing vehicles are too sparse or their moving direction is complex, our system performs poorly. To overcome this problem, lane detection and segmentation can be used to reduce dependence on traffic flow during future work.

5. Conclusions

This paper introduces a novel UAV-assisted wide-area multi-camera space alignment approach based on a spatiotemporal feature map. The proposed methods contains two key parts: spatiotemporal feature map construction and cross-view space matching. The first is presented on the basis of motion consistency between UAV-assisted aerial data and ground monitoring data. Following the procedure of feature line detection, spatiotemporal information extraction and feature map description, all input monitoring videos are described by spatiotemporal feature maps. The second key module is the cross-view space matching strategy, which is proposed to find the corresponding relationships between aerial and ground data. Through three matching steps, which are global feature map matching, air-to-ground time synchronization and cross-view spatial alignment, we can obtain a set of air-to-ground corresponding pixel pairs. In this way, the spatial relationship between assisted UAV and ground deployed camera can be calculated. Due to the united coordinates between UAVs, multiple cameras are successfully aligned into one coordinated system with UAV assistance.

Experimental results on simulation environment and real scene demonstrate that our system achieves satisfactory performance and aligns multiple camera in one space coordinate system. From the quantitative analysis, its minimum pixel error is around 5 pixels and the maximum error is less than 25 pixels. Through parameter discussion, we find that high deployment height and large pitch angle of camera are beneficial to alignment accuracy. Meanwhile, the proposed method shows superior performance to other contrast methods. Furthermore, this study has great academic meaning for camera pose estimation, camera array imaging and cross-camera information fusion. It has significant application value in the field of traffic monitoring, public security and so on. However, there may be some possible limitations to this study. The proposed method cannot work in no UAV fly zones which cannot obtain UAV-assisted data. Because the proposed method relies on traffic flow, it not applicable to the area with not enough traffic. Our future work will consider these problems.

Author Contributions: Conceptualization, J.L., Y.X., C.L. and T.Y.; Data curation, J.M. and Z.D.; Formal analysis, C.L.; Funding acquisition, J.L.; Methodology, J.L., Y.X. and T.Y.; Resources, Y.X.; Validation, J.M. and Z.D.; Writing—original draft, C.L.; and Writing—review and editing, J.L., Y.D. and T.Y. All authors have read and agreed to the published version of the manuscript.

Funding: This research was funded by National Natural Science of China (Nos. 62073262 and 61672429), Key Research and Development Program of Shaanxi (No. S2021-YF-ZDCXL-ZDLGY-0127), the Fundamental Research Funds for the Central Universities and the Innovation Fund of Xidian University (No. 20109205456).

Institutional Review Board Statement: Not applicable.

Informed Consent Statement: Not applicable.

Data Availability Statement: The data and the details regarding where data supporting reported results in this paper are available from the corresponding author.

Conflicts of Interest: The authors declare no conflict of interest.

References

1. Tang, Z.; Naphade, M.; Liu, M.; Yang, X.; Birchfield, S.; Wang, S.; Kumar, R.; Anastasiu, D.C.; Hwang, J. CityFlow: A City-Scale Benchmark for Multi-Target Multi-Camera Vehicle Tracking and Re-Identification. In Proceedings of the IEEE Conference on Computer Vision and Pattern Recognition, Long Beach, CA, USA, 15–20 June 2019; pp. 8797–8806.
2. Yang, T.; Li, Z.; Zhang, F.; Xie, B.; Li, J.; Liu, L. Panoramic UAV Surveillance and Recycling System Based on Structure-Free Camera Array. *IEEE Access* **2019**, *7*, 25763–25778. [CrossRef]
3. Deng, H.; Fu, Q.; Quan, Q.; Yang, K.; Cai, K. Indoor Multi-Camera-Based Testbed for 3-D Tracking and Control of UAVs. *IEEE Trans. Instrum. Meas.* **2020**, *69*, 3139–3156. [CrossRef]
4. Yang, T.; Ren, Q.; Zhang, F.; Xie, B.; Ren, H.; Li, J.; Zhang, Y. Hybrid Camera Array-Based UAV Auto-Landing on Moving UGV in GPS-Denied Environment. *Remote Sens.* **2018**, *10*, 1829. [CrossRef]
5. Hsu, H.; Wang, Y.; Hwang, J. Traffic-Aware Multi-Camera Tracking of Vehicles Based on ReID and Camera Link Model. In Proceedings of the 28th ACM International Conference on Multimedia, Seattle, WA, USA, 12–16 October 2020; pp. 964–972.
6. Cai, W.; Yang, J.; Yu, Y.; Song, Y.; Zhou, T.; Qin, J. PSO-ELM: A Hybrid Learning Model for Short-Term Traffic Flow Forecasting. *IEEE Access* **2020**, *8*, 6505–6514. [CrossRef]
7. Truong, A.M.; Philips, W.; Deligiannis, N.; Abrahamyan, L.; Guan, J. Automatic Multi-Camera Extrinsic Parameter Calibration Based on Pedestrian Torsors †. *Sensors* **2019**, *19*, 4989. [CrossRef]
8. Khoramshahi, E.; Campos, M.B.; Tommaselli, A.M.G.; Vilijanen, N.; Mielonen, T.; Kaartinen, H.; Kukko, A.; Honkavaara, E. Accurate Calibration Scheme for a Multi-Camera Mobile Mapping System. *Remote Sens.* **2019**, *11*, 2778. [CrossRef]
9. Yin, L.; Luo, B.; Wang, W.; Yu, H.; Wang, C.; Li, C. CoMask: Corresponding Mask-Based End-to-End Extrinsic Calibration of the Camera and LiDAR. *Remote Sens.* **2020**, *12*, 1925. [CrossRef]
10. Castanheira, D.; Silva, A.; Gameiro, A. Set Optimization for Efficient Interference Alignment in Heterogeneous Networks. *IEEE Trans. Wirel. Commun.* **2014**, *13*, 5648–5660. [CrossRef]
11. Lv, F.; Zhao, T.; Nevatia, R. Camera Calibration from Video of a Walking Human. *IEEE Trans. Pattern Anal. Mach. Intell.* **2006**, *28*, 1513–1518. [PubMed]
12. Liu, J.; Collins, R.; Liu, Y. Surveillance Camera Autocalibration based on Pedestrian Height Distributions. In Proceedings of the British Machine Vision Conference, Dundee, UK, 29 August–2 September 2011; pp. 1–11.
13. Liu, J.; Collins, R.T.; Liu, Y. Robust Autocalibration for A Surveillance Camera Network. In Proceedings of the 2013 IEEE Workshop on Applications of Computer Vision, Clearwater Beach, FL, USA, 15–17 January 2013; pp. 433–440.
14. Bhardwaj, R.; Tummala, G.K.; Ramalingam, G.; Ramjee, R.; Sinha, P. AutoCalib: Automatic Traffic Camera Calibration at Scale. *ACM Trans. Sens. Netw.* **2018**, *14*, 19:1–19:27. [CrossRef]
15. Wu, F.; Hu, Z.; Zhu, H. Camera Calibration with Moving One-dimensional Objects. *Pattern Recognit.* **2005**, *38*, 755–765. [CrossRef]
16. Zhang, Z. A Flexible New Technique for Camera Calibration. *IEEE Trans. Pattern Anal. Mach. Intell.* **2000**, *22*, 1330–1334. [CrossRef]
17. Abdel-Aziz, Y.I.; Karara, H.M. Direct Linear Transformation from Comparator Coordinates into Object Space Coordinates in Close-Range Photogrammetry. *Photogramm. Eng. Remote Sens.* **2015**, *81*, 103–107. [CrossRef]
18. Marcon, M.; Sarti, A.; Tubaro, S. Multi-camera Rig Calibration by Double-sided Thick Checkerboard. *IET Comput. Vis.* **2017**, *11*, 448–454. [CrossRef]

19. Unterberger, A.; Menser, J.; Kempf, A.; Mohri, K. Evolutionary Camera Pose Estimation of a Multi-Camera Setup for Computed Tomography. In Proceedings of the IEEE International Conference on Image Processing, Taipei, Taiwan, 22–25 September 2019; pp. 464–468.
20. Huang, L.; Da, F.; Gai, S. Research on Multi-camera Calibration and Point Cloud Correction Method based on Three-dimensional Calibration Object. *Opt. Lasers Eng.* **2019**, *115*, 32–41. [CrossRef]
21. Yin, H.; Ma, Z.; Zhong, M.; Wu, K.; Wei, Y.; Guo, J.; Huang, B. SLAM-Based Self-Calibration of a Binocular Stereo Vision Rig in Real-Time. *Sensors* **2020**, *20*, 621. [CrossRef] [PubMed]
22. Mingchi, F.; Panpan, J.; Yibo, L.; Jingshu, W. Research on Calibration Method of Multi-camera System without Overlapping Fields of View Based on SLAM. *J. Phys. Conf. Ser.* **2020**, *1544*, 012047.
23. Xu, Y.; Gao, F.; Zhang, Z.; Jiang, X. A Calibration Method for Non-overlapping Cameras based on Mirrored Absolute Phase Target. *Int. J. Adv. Manuf. Technol.* **2019**, *104*, 9–15. [CrossRef]
24. Mingchi, F.; Shuai, H.; Jingshu, W.; Bin, Y.; Taixiong, Z. Accurate Calibration of A Multi-camera System Based on Flat Refractive Geometry. *Appl. Opt.* **2017**, *56*, 9724.
25. Sarmadi, H.; Mu noz-Salinas, R.; Berbís, M.Á.; Carnicer, R.M. Simultaneous Multi-View Camera Pose Estimation and Object Tracking With Squared Planar Markers. *IEEE Access* **2019**, *7*, 22927–22940. [CrossRef]
26. Van Crombrugge, I.; Penne, R.; Vanlanduit, S. Extrinsic Camera Calibration for Non-overlapping Cameras with Gray Code Projection. *Opt. Lasers Eng.* **2020**, *134*, 106305. [CrossRef]
27. Yin, L.; Wang, X.; Ni, Y.; Zhou, K.; Zhang, J. Extrinsic Parameters Calibration Method of Cameras with Non-Overlapping Fields of View in Airborne Remote Sensing. *Remote Sens.* **2018**, *10*, 1298. [CrossRef]
28. Jeong, J.; Cho, Y.; Kim, A. The Road is Enough! Extrinsic Calibration of Non-overlapping Stereo Camera and LiDAR using Road Information. *IEEE Robot. Autom. Lett.* **2019**, *4*, 2831–2838. [CrossRef]
29. Dubská, M.; Herout, A.; Juránek, R.; Sochor, J. Fully Automatic Roadside Camera Calibration for Traffic Surveillance. *IEEE Trans. Intell. Transp. Syst.* **2015**, *16*, 1162–1171. [CrossRef]
30. Cobos, M.; Antonacci, F.; Comanducci, L.; Sarti, A. Frequency-Sliding Generalized Cross-Correlation: A Sub-Band Time Delay Estimation Approach. *IEEE ACM Trans. Audio Speech Lang. Process.* **2020**, *28*, 1270–1281. [CrossRef]
31. Berndt, D.J.; Clifford, J. Using Dynamic Time Warping to Find Patterns in Time Series. In Proceedings of the AAAI Workshop on Knowledge Discovery in Databases, Seattle, WA, USA, 31 July 1994; pp. 359–370.
32. Shah, S.; Dey, D.; Lovett, C.; Kapoor, A. AirSim: High-Fidelity Visual and Physical Simulation for Autonomous Vehicles. In Proceedings of the International Conference on Field and Service Robotics, Zurich, Switzerland, 12–15 September 2017; Volume 5, pp. 621–635.
33. Lowe, D.G. Distinctive Image Features from Scale-Invariant Keypoints. *Int. J. Comput. Vis.* **2004**, *60*, 91–110. [CrossRef]
34. Sarlin, P.; DeTone, D.; Malisiewicz, T.; Rabinovich, A. SuperGlue: Learning Feature Matching With Graph Neural Networks. In Proceedings of the IEEE Conference on Computer Vision and Pattern Recognition, Seattle, WA, USA, 14–19 June 2020; pp. 4937–4946.
35. DeTone, D.; Malisiewicz, T.; Rabinovich, A. SuperPoint: Self-Supervised Interest Point Detection and Description. In Proceedings of the IEEE Conference on Computer Vision and Pattern Recognition Workshops, Salt Lake City, UT, USA, 18–22 June 2018; pp. 224–236.
36. Schönberger, J.L.; Frahm, J.M. Structure-from-Motion Revisited. In Proceedings of the IEEE Conference on Computer Vision and Pattern Recognition, Las Vegas, NV, USA, 27–30 June 2016; pp. 4104–4113.
37. Schönberger, J.L.; Zheng, E.; Pollefeys, M.; Frahm, J.M. Pixelwise View Selection for Unstructured Multi-View Stereo. In Proceedings of the European Conference on Computer Vision, Amsterdam, The Netherlands, 11–14 October 2016; pp. 501–518.
38. Brahmbhatt, S.; Gu, J.; Kim, K.; Hays, J.; Kautz, J. Geometry-Aware Learning of Maps for Camera Localization. In Proceedings of the IEEE Conference on Computer Vision and Pattern Recognition, Salt Lake City, UT, USA, 18–22 June 2018; pp. 2616–2625.

MDPI

St. Alban-Anlage 66

4052 Basel

Switzerland

Tel. +41 61 683 77 34

Fax +41 61 302 89 18

www.mdpi.com

Remote Sensing Editorial Office

E-mail: remotesensing@mdpi.com

www.mdpi.com/journal/remotesensing

www.ingramcontent.com/pod-product-compliance
Lightning Source LLC
LaVergne TN
LVHW070852170726
843515LV00005B/626